D1499235

3 76 · 1
21

Forthcoming titles

Outline

Editors

George M. Dunnet
*Regius Professor of Natural History,
University of Aberdeen*

Charles H. Gimingham
*Professor of Botany,
University of Aberdeen*

Editors' Foreword

Both in its theoretical and applied aspects, ecology is developing rapidly. This is partly because it offers a relatively new and fresh approach to biological enquiry; it also stems from the revolution in public attitudes towards the quality of the human environment and the conservation of nature. There are today more professional ecologists than ever before, and the number of students seeking courses in ecology remains high. In schools as well as universities the teaching of ecology is now widely accepted as an essential component of biological education, but it is only within the past quarter of a century that this has come about. In the same period, the journals devoted to publication of ecological research have expanded in number and size, and books on aspects of ecology appear in ever-increasing numbers.

These are indications of a healthy and vigorous condition, which is satisfactory not only in regard to the progress of biological science but also because of the vital importance of ecological understanding to the well-being of man. However, such rapid advances bring their problems. The subject develops so rapidly in scope, depth and relevance that text-books, or parts of them, soon become out-of-date or inappropriate for particular courses. The very width of the front across which the ecological approach is being applied to biological and environmental questions introduces difficulties: every teacher handles his subject in a different way and no two courses are identical in content.

This diversity, though stimulating and profitable, has the effect that no single text-book is likely to satisfy fully the needs of the student attending a course in ecology. Very often extracts from a wide range of books must be consulted, and while this may do no harm it is time-consuming and expensive. The present series has been designed to offer quite a large number of relatively small booklets, each on a restricted topic of fundamental importance which is likely to constitute a self-contained component of more comprehensive courses. A selection can then be made, at reasonable cost, of texts appropriate to particular courses or the interests of the reader. Each is written by an acknowledged expert in the subject, and is intended to offer an up-to-date, concise summary which will be of value to those engaged in teaching, research or applied ecology as well as to students.

Studies in Ecology

Vegetation Dynamics

JOHN MILES

Principal Scientific Officer,
Institute of Terrestrial Ecology,
Banchory, Kincardineshire

LONDON
CHAPMAN AND HALL
A Halsted Press Book, John Wiley & Sons, New York

First published 1979
by Chapman and Hall Ltd
11 New Fetter Lane, London EC4P 4EE
© *1979 J. Miles*
Printed in Great Britain at the
University Press, Cambridge

ISBN 0 412 15530 3

This paperback edition is sold subject to the condition that it shall not, by way of trade or otherwise, be lent, resold, hired out or otherwise circulated without the publisher's prior consent in any form of binding or cover other than that in which it is published and without a similar condition including this condition being imposed on the subsequent purchaser.

All rights reserved. No part of this book may be reprinted, or reproduced or utilized in any form or by any electronic, mechanical or other means, now known or hereafter invented, including photocopying and recording, or in any information storage and retrieval system, without permission in writing from the Publisher.

Distributed in the U.S.A.
by Halsted Press, a Division
of John Wiley & Sons, Inc., New York

Library of Congress Cataloging in Publication Data

Miles, J
  Vegetation dynamics.

  (Outline studies in ecology)
  "A Halsted Press book."
  Includes bibliographical references and index.
  1. Vegetation dynamics. I. Title. II. Series.
QK910.M55   1978   581.5   78-13070
ISBN 0-470-26504-3

90783   BELMONT COLLEGE LIBRARY

Contents

QK
910
.M55
1979

Preface

Vegetation dynamics is an important subject. A knowledge and understanding of it is central to the science of vegetation management–in grassland, range and nature reserve management, and in aspects of wildlife management, forestry and agricultural crop production. It is also a large and diffuse subject. In a small book such as this I had to be highly selective, and could not do equal justice to all aspects. I have had therefore to condense many examples, and more regrettably, many arguments. While I have tried to present a broad selection of topics and examples, the content inevitably reflects my own special interests and experience.

The study of vegetation and its dynamics does not lend itself to neat and tidy divisions, and the way of allotting material into different chapters here is arbitrary. I have used Chapter 1 to introduce a number of ideas, beginning with the nature of vegetation in space, then passing to an introduction to the nature of changes in vegetation with time, in particular those generally known as successions. The book also contains a number of asides to the text's central arguments; I hope the reader finds these interesting rather than disconcerting.

There is no easy way of learning about vegetation and its dynamics. A book like this can give generalizations and a few examples, but there are exceptions to most generalizations. Understanding and a feeling for the subject come only with increasing experience of vegetation in the field, and through reading as widely as possible in the relevant literature. I have therefore tried to give a generous number of references to source material, and of references recommended for further study.

I am grateful to those authors and publishers who gave permission to reproduce material, and apologise where such material has, in the event, not been included.

July, 1978

J. M.

1 The nature of vegetation

1.1 Ubiquity of changes in time

Vegetation dynamics is the study of changes in vegetation with time. Such changes were recognized as long ago as ca. 300 B.C. by Theophrastus [1], while increasingly detailed and critical descriptions have been made over the last three centuries [2].

Change in vegetation is universal. Many examples are matters of common experience, such as the constant appearance of weeds in lawns and gardens, or the encroachment of scrub and thicket on an abandoned field or meadow. Others may be less obvious, either because change is slow in relation to the human life span, or because it is merely an internal re-ordering of pattern in a patch of vegetation that keeps the same overall appearance.

Any patch of vegetation is a dynamic thing. Individual plants of all species represented are born and die. If environmental conditions change, and this includes the ways in which plants themselves modify their environment, the balance between birth and death rates for particular species may change, so that the relative abundance of individuals of the different species also changes. If deaths exceed births for long enough, species will disappear. (Strictly, a species cannot disappear from a patch, only the individuals present that are classified within that particular species category. However, the correct form of words is so cumbersome in repetitious use that the abbreviated form will be used from now on.)

The patch is subjected to a yearly 'rain' of reproductive units from the surrounding vegetation. This is usually termed a seed rain, though it also comprises fruits and asexually produced propagules. The patch will tend to 'resist' these invasions, especially where there is complete cover. Most incoming propagules will fail to establish, though a few may, and new species can appear in the patch in this manner. The margins are also exposed to penetration by ramets of vegetatively expanding individuals or clones, and this is an important source of change in many kinds of vegetation. (These matters are dealt with more fully in Chapter 2.)

1.2 Variation in space

Any analysis of the nature of changes in vegetation with time must first consider the nature of vegetation in space, that is, as it exists over the ground at any one time. This is fundamental. A researcher's basic attitudes and assumptions about vegetation govern his objectives and approach to any problem. These predispose him towards using particular methods and techniques, and thus to collecting particular kinds of data with their own special inherent limitations of interpretation. Even the

7

widespread use of the term 'plant community' to denote any grouping of plants on the ground worthy of study may cause unwitting bias. The term implies internal relationships, which are often unknown or arguably non-existent, and a certain spatial discreteness, which may be lacking [3, 4]. Despite this, the use of a less ambiguous term, such as assemblage [5], or patch, as preferred here, has never become popular.

How vegetation is regarded in fact depends on the relative emphasis given to its different observed properties, and to the relative importance or obviousness of these properties in vegetation at different places.

1.2.1 Emphasis on uniqueness

It is readily determinable in the field that no two patches of vegetation are ever exactly alike in the combinations and proportions of the different species present. Gleason pointed this out over fifty years ago [6], and it has been stressed periodically since [7, 8, 9]. Gleason put forward cogent arguments as to why this was inevitable [6, 8]. He maintained that the properties of vegetation depended completely on the properties of the individual plants composing it—the 'individualistic' concept of vegetation. Similar ideas were put forward independently at about the same time in France and Russia [10, 11, 12, 13]. The vegetation of an area was seen as the result of selection by an environment fluctuating in time and varying in space on plants arriving there by a similarly fluctuating and fortuitous process of immigration. There is no significant body of data existing which leads to the rejection of this view. (The environment is again taken to include the modifying effects of plants themselves.) In fact the role of chance in the development of vegetation is stressed repeatedly by examples in later chapters. The only major advance in our overall understanding since Gleason's day is that it is now possible to attempt to treat many factors, including the chance processes of plant migration, in a probabilistic manner. Thus, though all patches of vegetation may be different and unique, it is possible to examine the likelihood of any patch, whether existing and fairly stable, actively changing, or newly forming, being more or less dissimilar to other known patches.

1.2.2 Emphasis on similarity

It is also true that patches of vegetation growing under similar environmental conditions and with similar histories of environment and plant migration are often very alike in composition (which Gleason also pointed out [8]). Thus clearly definable 'types' may be recognized, as indeed they have for thousands of years. This frequent similarity led to the idea of 'plant communities' which repeated themselves in space, and even led some workers to consider such a community as analogous to an organism, an idea advanced with especial vigour by Clements [2].

This extreme 'organismic' view is clearly untenable however. A patch of vegetation may be composed of species that have had very different evolutionary histories. It cannot be traced along a distinct ancestral phylogenetic line and so is fundamentally different from an organism [14, 15]. Fossil evidence has shown that present kinds of vegetation in Europe, for example, have no long history in the Quaternary but are just temporary aggregations resulting from particular environmental and historical factors. In the past, the component species of any vegetation type may have occurred in quite different combinations and proportions [16, 17, 18]. Vegetation thus cannot be logically classified on the basis of presumed similarity of genotype and evolutionary history in the same way as individual organisms.

However, classifications of vegetation are often needed. In any given region it is usually possible to distinguish a number of different types which have practical value in ecological research, soil surveying, and in land use fields such as site quality assessment in forestry, agriculture, and range and wildlife management. That such types generally either intergrade, and so must be arbitrarily divided, or else may represent only narrowly defined units to be used as reference points against which all 'intermediate' types can be set [19], does not prevent them from being useful. It is notable, though, that while many kinds of logical classification of vegetation can be devised, many workers still use a system analogous to that used in the taxonomy of individual organisms [20], with classes, orders and so on. Presumably the view of vegetation as an organism is not quite dead.

Notwithstanding these comments, a patch of vegetation does show, to a greater or lesser extent, the development of properties which are characteristic of the assemblage as a whole rather than of the component species and individuals. These include structural organization, both vertically (with different life forms present in distinct layers or strata) and horizontally (vegetation pattern), species diversity (and hence food chain diversity) and vegetation stability. While the assemblage is historically just a temporary aggregation, many of the component species may have been evolving in close proximity for long periods. Each of the component species is presumably still evolving under selection pressures that include the influence of the surrounding species. Therefore selection both affects and is a function of the vegetation as a whole (21). Species may develop adaptations such that they are substantially dependent on others for their niches, for example epiphytes, and the ground layer herbs of temperate deciduous woodland which flower before or during bud burst and leaf expansion in the tree layer above. Structural organization within vegetation may be negligible in pioneer assemblages such as the populations of ephemeral weeds on cultivated ground, but is very marked in forest, especially in tropical rain forest with its complex array of life forms and layers [22, 23]. Individuals within an assemblage may even be connected organically. Clonal development by rhizomes, root suckers, and so on is commonplace in herbs, frequent in shrubs and occasional in trees. Further, natural intraspecific root grafting is common in trees [24], and may also be in other life forms, and the effects of this can act to reduce the consequences of competition between individuals [25]. However, structual organization and other properties of vegetation are still functions of the species present, reflecting the cumulative effects of species evolution rather than real emergent properties of the vegetation *per se* [26,

9

27]. Nevertheless, any patch does represent a group of interacting and co-evolving species. In addition, the behaviour of a population growing in a mixture of species can differ markedly from that growing in a pure stand [28]. All in all, therefore, it seems purely a matter of personal taste and emphasis whether the properties of vegetation are considered to reflect just the sum of the properties of the component species, or whether vegetation is regarded as possessing its own emergent properties, perhaps becoming an 'emergent holistic system' (29).

1.2.3 Emphasis on intergradation

A clearly demarcated patch of vegetation in the field results from the area having an environmental history critically different from that surrounding it. However, while environmental and vegetational changes may be fairly abrupt, as at a beach strand line or the edge of a swath of wind-felled trees, they are often more gradual. Examples include the changes seen with increasing altitude on a mountain, or those on certain coastal grasslands on the west of Scotland that are influenced by wind-blown sand largely composed of pulverized sea-shells, causing a gradient of decreasing calcium and pH away from the beach. Emphasizing such continuities or gradients, the way in which different vegetation patches intergrade or in which species and species assemblages are related along environmental gradients, gives a third major way of viewing the nature of vegetation. This method of study is usually termed gradient analysis. While this way of looking at vegetation was recognized at least sixty years ago [30], it has really only been developed over the last thirty years, particularly by Curtis and Whittaker in the United States [31, 32]. Thus Whittaker has demonstrated how species and species assemblages tend to show broadly overlapping population distributions along environmental gradients, mostly in the form of the binomial distribution of statisticians [32, 33]. When plotted and smoothed they tend to appear as bell-shaped curves (Fig. 1.1). These reflect the differing environmental tolerances of

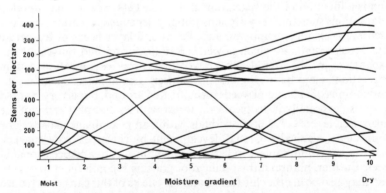

Fig. 1.1 Changes in abundance of different trees and shrubs (each line refers to a different species) in vegetation along topographic moisture gradients from (above) the Siskyou Mountains, Oregon, and (below) the Santa Catalina Mountains, Arizona. (From Whittaker 1967 [32], courtesy of Cambridge University Press).

10

individual species. From this view, vegetation is seen as a complex pattern of overlapping species populations, which was concisely described by Whittaker thus [32]: 'Numerous plant species, with population centres scattered along environmental gradients, each with binomial distributions broadly overlapping those of other species, freely and variously combine into communities which predominantly intergrade with one another, forming a complex and potentially continuous but variously interrupted population pattern.'

1.2.4 Conclusions

Vegetation hardly needs defining. It is the carpet of plant life covering the ground, whether of a square metre or an entire continent. It is a unique kind of living system. Its properties are only those of its component species, yet these are modified within the vegetation by interspecific competition. It can show a complex structural organization and spatial arrangement of species, which to some extent can be repeated from place to place, yet the majority of these species interrelations are probably facultative, not obligatory. It can be viewed as a mosaic of distinct patches or types, or as a pattern of intergrading populations. All patches are different and unique, yet some are more similar than others such that vegetation can be classified into different types, which are of great practical utility. All these views are demonstrably justifiable. It is thus a matter of judgement which is better suited to any particular task or research problem. Gradient analysis has proved very useful in studying the ecology of species, and it and classification have furthered our understanding of vegetation relationships and variation. In this book attention is focused variously at the level of the species and the arbitrarily delimited patch or vegetation type, and on gradients in time and between different kinds of change.

1.3 The nature of vegetation in time

The study of vegetation change has unfortunately developed a particularly rich and confusing jargon [34], in which the same term has sometimes been used in different ways and different terms have been used for the same phenomenon. Mostly however, a distinction has been made between changes in vegetation that keep the same overall appearance, termed *fluctuations*, and changes which markedly alter the appearance of a patch such that it can be considered to have changed into a different type, termed *succession*. Succession depends on changes in those species that give a specific form, colour or other characteristic to a patch of vegetation, through which it is distinguishable by eye from an adjacent patch. These may be termed *primary* species, the others, *secondary* species [5]. Both kinds of change result from the fact pointed out at the beginning of this chapter, that all vegetation is in a continuous state of flux, with individuals dying from old age or other causes and being replaced by new individuals of the same or other species.

Fluctuational changes in vegetation result from differences in the seasonal growth cycles of the component species, and from the constant re-ordering of internal pattern following normal birth and death processes. Year to year variations in climatic or other environmental factors influence these processes, and can result in quite marked differences in the proportions of the different species present, without markedly altering the overall importance of the primary species. Grasslands are a prime example of this. Successional changes may be seen as a more extreme form of this kind of fluctuation, to the extent that the combinations and proportions of the primary species present do change. It should be stressed that the distinction between fluctuations and successions is made for convenience, and if phenological differences are excluded, is quite arbitrary. The common underlying processes of plant population change differ only in degree.

Fluctuations occur in all vegetation, though their extent varies greatly between vegetation types. All vegetation appears to be also subject to successional change, though this may often be too slow to be detectable. Fluctuational changes may be thought of as short term and reversible, with the vegetation varying from year to year (or period of years) about a notional mean, and typically not involving the appearance of new species [35]. In contrast, successional changes are directional, with the vegetation composition tending away from the initial notional mean, and frequently though not necessarily involving the establishment of new species. Successional change can occur over almost any time scale, from the few months needed to establish a sward of mosses or ephemerals on bared soil to the changes over the thousands and millions of years of long term climatic change and of evolutionary and geological time. (Many workers prefer not to label these latter changes as successions, but there seems no logical cut-off point. Natural selection and evolution may be apparent over short as well as long time scales [36].)

Fluctuational changes in vegetation are less dramatic than those of successions. In temperate climates they are less important to people like farmers or range managers who try to control the composition of swards, although in arid climates, fluctuations following variations in rainfall can be of critical importance. By and large they have been less studied than successions, and are given correspondingly less attention here.

In passing, it may have been noticed that in this section I have adopted a working definition of a vegetation type as one characterized by the particular combination and relative abundance of its primary species, i.e. those that are physiognomically dominant or numerically predominant. This is at the expense of any differences between types in numerically unimportant species of narrow ecological tolerance, which might in fact be much more sensitive in reflecting differences in particular critical environmental factors. There are two reasons for this choice. Firstly, it is the most abundant species in vegetation which have the greatest influence on the other species present and on the soil. They also govern the flammability, and the attractiveness to grazing animals, of the vegetation, both of which can have profound effects. In ecological terms they are the most important species. Secondly, a vegetation type defined by its primary species can be recognized at a glance, an important need (just as it is that a species, to be useful to an ecologist working in the field, must also be immediately recognizable).

1.3.1 Development of awareness of change

Awareness of vegetation succession became widespread in the last century among botanists and foresters [2, 37]. For a long time though, succession seems to have been regarded as the exception rather than the rule, and the idea that it was a universal process seems to have arisen towards the end of the century from the work of Hult and of Warming [2]. An awareness of fluctuations, which are anyway less exciting to most researchers, seems only gradually to have arisen during this century. Following Warming, the idea of succession was developed particularly in the United States, first by Cowles [38, 39, 40] and then by Clements [2, 41, 42].

1.3.2 Models of succession

In 1916, Clements produced a monumental and brilliant summary and codification of his ideas on succession [2], which quickly became so well known that it is with him that the concept has come to be most popularly associated. His model was comprehensive and intellectually satisfying. This is probably a major reason why it has persisted for so long, even though it was immediately noted, in particular by Gleason [43, 44], that it was not consistent with all the facts. Its enduring influence is shown by the fairly uncritical summaries of it even in recent textbooks [45, 46, 47, 48], though with a glowing exception by Colinvaux [49].

In brief, Clements saw succession as a directional change of vegetation types, each successive type establishing itself because the preceding type had modified the site in a way favourable to its successor, this sequence finally ending in a *climax* type that was stable and self-maintaining under the current conditions of site and climate (Fig. 1.2). This was seen as a deterministic process. If the climax was destroyed by some catastrophic event, it was believed its preceding stages would be repeated and the same climax eventually re-established provided the climate was unchanged.

There are numerous objections to this model. In particular, clear-cut stages are usually absent. Any 'stage' has to be arbitrarily defined. Insofar as any appears, it is a result of gradual population changes, the equivalent in time of the population gradients in space demonstrated by Whittaker (Fig. 1.1). Secondly, although the site is modified during the succession, Clements overemphasized the importance of these changes in facilitating

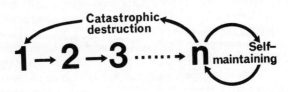

Fig. 1.2 Schematic outline of Clements' model of succession as a deterministic process with separate stages leading to a self-maintaining climax (n).

13

the establishment of species which predominate later in the succession. Egler and others [1, 50, 264, 287] have stressed that most species noted during successions on abandoned fields (often termed old-field successions), or when vegetation is disturbed, were in fact present at the outset as buried seed, roots, rhizomes etc, or invaded very shortly after. Subsequent changes in the vegetation were due to different rates of growth, reproduction and survival of the species present. 'The annual weeds with their rank growth are first in evidence, with other plants occurring as seeds or seedlings. The annual weeds drop out of the race, and grasses assume predominance, with woody plants present but small or dormant. As each successive group drops out, a new group of species, there from the start, assumes predominance. Eventually only the trees are left. Developments in non-forest regions have analogous series of stages.' Egler termed this the *initial floristic composition* factor. The species disappearing from the vegetation will usually persist for a time, varying from decades to centuries, as viable seed buried in the soil [51, 52]. This 'seed bank' ensures the importance of the initial floristic composition factor in successions on bared ground. (Clements noted that propagules of the dominants of all 'stages' might be present at the time of initiation of such successions, but seems to have overlooked the wider significance of this.) A further objection to the classical model is that as the composition of no two patches of vegetation is precisely the same, neither are the seed banks. Succession on different patches of disturbed ground in the same locality frequently proceeds quite differently because of such differences. Variation in seed banks is illustrated in Table 1.1. This gives estimates from nine heather moorland sites in the Scottish Highlands. (These are sites dominated by the dwarf shrub *Calluna vulgaris*, created by persistent fire and grazing in past centuries following destruction of the original woodland cover, and since maintained in a remarkably stable state by regular burning.) Although all the seed populations show a predominance of *Calluna*, the combinations and proportions of the other species present vary markedly from site to site.

Succession on disturbed ground, termed *secondary* succession, as opposed to colonization and succession on virgin surfaces, termed *primary* succession, is thus demonstrably not a deterministic but a probabilistic process. This also seems to be true of primary successions, though much less is known about them. Acceptance of Gleason's ideas about the nature of vegetation (Section 1.2.1) also leads to rejection of any idea of a widespread climatic climax. Even before Clements published his monograph, Harper [53] had noted that every different soil type seemed to have its own characteristic successional and climax vegetation. Also, present knowledge of past variations in climate suggests that it will vary to a similar extent in the future. This must throw doubt on the probability of any really long term stability of vegetation. Further, the vegetation in many parts of the world appears anyway to be in a state of gross flux, with natural catastrophic events like drought, fire and hurricanes preventing stability on a time scale longer than the lifespan of the primary species, for example the grasslands of the Great Plains [54, 55], and much of the

Table 1.1 Estimated numbers per 600 cm² of viable seeds buried in the topsoil at nine sites with dominant *Calluna vulgaris* in the Scottish Highlands

	1	2	3	4	5	6	7	8	9
Agrostis canina	2			13	12	13	3		19
A. tenuis			1			73	8		14
Anenome nemorosa									1
Anthoxanthum odoratum					2	8			
Aphanes arvensis						1			
Betula pendula	1	1	2	1			1	133	
B. pubescens			11			3			
Calluna vulgaris	881	800	742	1258	651	510	484	516	464
Carex spp.	26	5	11	28	49	10	11		40
Deschampsia flexuosa					1		1	6	11
Empetrum nigrum			26						
Erica cinerea			1			43	10		
E. tetralix				93	390		46		
Festuca rubra						11			1
Galium saxatile				3		7			5
Genista anglica						1			2
Holcus lanatus					1	1	1		
Hypericum pulchrum					12	1	1		18
Juncus bulbosus	1	2		79	74			3	5
J. effusus									4
J. squarrosus	1	10	4	39	56	1	4		
Lotus corniculatus	1								
Luzula multiflora	4	4	1	1	4	8	12		4
Narthecium ossifragum					1				
Polygala serpyllifolia		1		1	4				
Potentilla erecta	1	1	1	1	5	2	8		30
Prunella vulgaris						1			
Rumex acetosa						1			
R. acetosella				1					
Sagina procumbens						12			
Sarothamnus scoparius						14			
Teucrium scorodonia								5	
Trichophorum cespitosum		3		3	4	1			
Vaccinium myrtillus			6						
Veronica officinalis						1		3	
V. serpyllifolia						3			6

natural forest of northern North America [56, 57, 58]. These points, coupled with the fact that vegetation not markedly influenced and modified by man's activities is becoming increasingly rare worldwide, make the concept of the 'natural' climax arguably redundant. It is useful to think of a 'terminal' vegetation type which would tend to be stable in relation to the human life span at any site under a given set of management conditions, but this is not climax.

It now seems that no single model can account for all successions. Clements' model (which has been termed the *facilitation* model [59] because of the stress laid on site modification as a driving force, or *relay floristics* [50] after the stage by stage process envisaged) may operate in some situations, though almost certainly not exclusively. While clear-cut 'stages' are arguably almost never found, site modification may be necessary in certain primary and secondary successions before particular species can establish (see Chapters 5 and 6). Egler's initial floristic composition factor seems to operate to a greater or lesser extent in all situations.

Once a vegetation cover has been achieved, the opposite of the facilitation model often seems to occur. That is, a population at any point along a successional sequence (termed a *sere* by Clements) may inhibit the invasion of other species by a variety of means, though perhaps most commonly by simple physical occupancy alone, which in particular means monopolization of light. This has been termed the *inhibition* model [59]. Its operation does not of course preclude the operation of the facilitation model at an earlier or later point in the succession. In fact a species which because of its presence may be inhibiting the establishment of another, may simultaneously be modifying the site in such a way that on its death a new species may establish which could not have done so before.

The current absence of one all-embracing model of succession to replace that of Clements may seem disappointing. However, given the complex and varied nature of vegetation, and the essential validity of Gleason's 'individualistic' concept, it would be surprising if patterns and mechanisms of succession were not as varied as the differences between individual species in dispersal efficiency, in ability to persist as seeds, and in ability to establish, grow, compete and reproduce at different sites and in different assemblages. Any given succession is the resultant of a large number of probabilities.

A basis for the assertions made in this section will be found in the following chapters. Detailed critiques of Clements' model have been made by McCormick [60] and by Drury and Nisbet [1, 61], and the reader is referred to these for a more complete account. However, excessive criticism of Clements is invidious. His analysis of the processes of succession was masterly, and will be used here as the framework for Chapter 2. Further, he was largely responsible for injecting the dynamic principle into a static concept of ecology in Britain and Europe. In Britain at least, this may still not have reached its entire audience, judging from the scanty information on successions and successional relationships in certain recent books dealing with vegetation (e.g. [62]).

1.3.3 Properties of successions
The model of succession developed by Clements is inextricably linked with his idea of the plant 'community' as an organism. 'The developmental study of vegetation necessarily rests upon the assumption that the unit for climax formation is an organic entity. As an organism the formation arises, grows, matures, and dies.' '. . . Each climax formation is able to reproduce itself, repeating with essential fidelity the stages of its development.' We have seen that this view is untenable. Rather than being a highly organized society of plants, a patch of vegetation is an

aggregation, essentially temporary in nature, of independently behaving plant species occurring on a site where each can grow and happens to have arrived through the chance processes of dispersal.

However, the organismic analogy is seductive, and it appears to have been adopted, perhaps often unwittingly, by many followers of that popular research field of the last twenty years, the ecosystem study. (The term ecosystem was coined by Tansley [63] to denote the total grouping of vegetation, animals, soil and climate of a site. Like vegetation it can cover any area, from a rotting tree stump to the entire planet.) The very term 'ecosystem' may in part have been responsible for this. A 'system' in its primary meaning is an organized or connected group of objects (an 'organized whole' in the original Greek), so the use of 'ecosystem' may lead the unwary into the organismic fallacy even more directly than the term 'community' which was criticized earlier.

A preoccupation with ecosystems as a research subject led to a predisposition to studying ecosystem 'properties' such as primary production, biomass accumulation, nutrient and carbon cycling, species diversity, and stability, in an attempt to develop a general theory of communities in terms of these functional and structural characteristics. Equally inevitably, attempts were made, in particular by Margalef [64, 65] and Odum [66], to develop a similar functional theory of succession by determining how such community and ecosystem properties varied with time during successions. The result has been summarized by Odum [66] as follows:

'Ecological succession may be defined in terms of the following three parameters. (i) It is an orderly process of community development that is reasonably directional, and, therefore, predictable. (ii) It results from modification of the physical environment by the community; that is, succession is community-controlled even though the physical environment determines the pattern, the rate of change, and often sets limits as to how far development can go. (iii) It culminates in a stabilized ecosystem in which maximum biomass (or high information content) and symbiotic function between organisms are maintained per unit of energy flow. In a word, the 'strategy' of succession as a short-term process is basically the same as the 'strategy' of long-term evolutionary development of the biosphere—namely, increased control of, or homeostasis with, the physical environment in the sense of achieving maximum protection from its perturbations.'

Trends which Odum believed may be expected during ecosystem development and succession are listed in Table 1.2. To this list Whittaker [33] added the further generalization that soil depth, organic matter content and horizon differentiation all tended to increase also with succession.

This 'contemporary' model, with its emphasis on predictability, community-control of change, and progression to a stable end-point, clearly seems neo-Clementsian, and has been devastatingly criticized by Drury and Nisbet [1]. They argue that there are so many exceptions to the

Table 1.2 A tabular model of ecological succession: trends to be expected in the development of ecosystems. (After Odum 1969 [66]. Copyright 1969 by the American Association for the Advancement of Science.)

Ecosystem attributes	Developmental stages	Mature stages
Community energetics		
1. Gross production/community respiration ratio	> or < 1	Approaches unity
2. Gross production/biomass ratio	High	Low
3. Biomass supported/unit energy flow ratio	Low	High
4. Net community production	High	Low
5. Food chains	Simple, linear	Complex, weblike
Community structure		
6. Total organic matter	Small	Large
7. Inorganic nutrients	Mainly in soil minerals	Mainly in organic matter
8. Species diversity–variety component	Low	High
9. Species diversity–evenness component	Low	High
10. Biochemical diversity	Low	High
11. Stratification and pattern diversity	Poorly organized	Well organized
Life history		
12. Niche specialization	Broad	Narrow
13. Size of organism	Small	Large
14. Life cycles	Short, simple	Long, complex
Nutrient cycling		
15. Mineral cycles	Open	Closed
16. Nutrient exchange rate between organisms and environment	Rapid	Slow
17. Role of detritus in nutrient regeneration	Unimportant	Important
Selection pressure		
18. Growth form	For rapid growth ('*r*-selection')	For feedback Control ('*K*-selection')
19. Production	Quantity	Quality
Overall homeostasis		
20. Internal symbiosis	Undeveloped	Developed
21. Nutrient conservation	Poor	Good
22. Stability (resistance to external perturbations)	Poor	Good
23. Entropy	High	Low
24. Information	Low	High

generalizations of the model that it should be rejected. They further note, as have Whittaker and Woodwell [26], that all of the so-called community or ecosystem 'properties' are functions of the species present. While most of the ecosystem properties listed in Table 1.2 are generally accepted, at least for successions leading to forest, a few are doubtful or arguably incorrect for many situations, like the shift in the nutrient pool (No. 7) and increasing stability (No. 22). Others are trifling. Thus, any growth of vegetation on bare soil must decrease entropy (i.e. increase molecular order) and increase 'information' (also basically 'order') (Nos. 23, 24) in the sense of Margalef [64]. Again, it is inevitable that as more species colonize a site during succession (rare species will tend to colonize later than common species, as propagates and scarce, and the probability of early arrival lower), niches must contract.

Drury and Nisbet propose, as an alternative, that most of the phenomena of succession can be understood as consequences of differential colonizing ability, growth and survival of species adapted to growth under different environments i.e. species with different physiological and ecological tolerances. Further:

> 'The appearance of successive replacement of one 'community' or 'association' by another results in part from interspecific competition which permits one group of plants temporarily to suppress more slowly growing successors. The structural and functional changes associated with successional change result primarily from the known correlations in plants between size, longevity, and slow growth. A comprehensive theory of succession should be sought at the organismic or cellular level, and not in emergent properties of communities.'

1.3.4 Sources of information on vegetation change

There are four main sources of evidence for and about changes in vegetation:

1. From direct observations through time.
2. From historical evidence.
3. From preserved biological evidence.
4. From examination of spatial sequences on adjacent sites which are believed to represent different points in the same succession.

Evidence from direct observation in time is common, though most has come from simply monitoring changes rather than inducing and measuring change under controlled experimental conditions. Almost all evidence for fluctuations, and much of that for secondary successions has been obtained in this way, but very little has been obtained for primary successions, and then only the early stages have usually been observed.

Historical evidence is generally rather scanty, usually does not go very far back in time, and is frequently of uneven or unassessable quality. However, written or cartographic records can give valuable information if cautiously interpreted [55, 67, 68]. Photographic evidence often falls into this category, and is unequivocal. An outstanding example is Hastings and Turner's [69] study of change in the arid Southwest of the United States and the arid Northwest of Mexico.

There are several kinds of preserved biological material that give useful evidence. The most well-known and well-worked source is from pollen preserved in stratigraphic sequences in peats and lake deposits [18, 70, 71]. Pollen buried in soils can also yield useful information about past vegetation, but as this at best only represents a quasi-stratigraphical sequence, with the oldest deposited pollen having been washed furthest down the profile, its interpretation is much more difficult and needs great caution [72, 73]. Charcoal fragments in lake sediments [74] or soils [58, 75], and opal phytoliths [76], woody stem fragments [58] and viable seed [52, 77] buried in soils can all also be useful sources of additional evidence.

Evidence from examination of existing spatial sequences of vegetation has been widely collected in the past, and is probably the main source of putative information about primary successions. These latter include the classic studies on the Lake Michigan sand dunes by Cowles [38] and Olson [78], on recessional glacial moraines in Alaska by Cooper [79, 80, 81] and later workers (though Cooper also monitored changes in permanent quadrats), and on submerged soils (*hydroseres*) in Britain by workers like Pearsall and Godwin, which formed the basis of Tansley's classic synthesis [82] of the hydrosere. However, such evidence is intrinsically the most potentially misleading source of all. The interpretation of spatial sequences of vegetation on different sites as though they represented changes in time on a single site depends for its validity on all sites having effectively identical soils and climatic and microclimatic conditions, and histories of environment and vegetation succession. Antecedent similarity can rarely be demonstrated satisfactorily, and in fact little evidence for it has usually been produced in past studies. Inevitably, misinterpretation occurs. Thus, a remarkable study by Walker of post-glacial hydroseres in Britain [83] has shown an unforeseen variety in the sequence of translations between 'stages', and that, far from woodland being the natural culmination of succession in the mires and fens over much of Britain as predicted by Tansley [82], it is likely to be *Sphagnum* bog!

2 Processes of vegetation change

Clements [2] recognized the following subprocesses of succession: (1) initiation, (2) immigration of new species, (3) establishment (or ecesis), (4) competition, (5) site modification (or reaction), (6) stabilization at the 'climax'. With the probable exception of stabilization, and a considerable reduction in emphasis of the role of site modification, which Clements saw as the main driving force of a succession after its initiation, this is still a useful way of subdividing the complex process of succession for study.

2.1 Initiation of successions and fluctuations

We saw in the previous chapter that all vegetation is in a continuous state of flux as individual plant units die and are replaced. The rate of species turnover of course varies with the life span of the species concerned. This may range from a few months for ephemerals like winter annuals [84] to 4,000 years or more for bristlecone pines (*Pinus aristata*) [85]. Apart from this natural flux, any patch of vegetation may be in a state of relative stability or of active change (instability). Successional or fluctuational change may be begun in a stable patch, or the direction of an existing change altered, if the environment of the patch alters. Three main environmental processes may be recognized: physiographic (including soil processes), climatic, and biotic. Changes in environmental factors may kill plants in the patch, and thus create niches for colonization by other species, whether already present in the patch or entering via the seed rain, or they may simply alter the balance of competition between the existing species such that some expand the size of individuals or of their populations at the expense of others.

2.1.1 Physiographic processes

Physiographic (or geomorphic) processes include: coastal [86] and other soil erosion by water; soil movements due to gravity e.g. soil creep, landslides, avalanches, scree movements and rockfalls on cliffs [87]; meandering of rivers [88, 89]; silting of lakes and estuaries [83, 90]; deposition of material by glaciers [79, 91]; volcanic eruptions [92]; flooding [5, 93]; soil erosion and redeposition by wind [78]; changes in soil stability and drainage caused by frost [94, 95].

All these processes either destroy old surfaces and vegetation, or provide new surfaces for colonization, or both. They may act in a gradual way, or suddenly and catastrophically, or both. Where new surfaces are produced, the ensuing successions have generally been considered as 'primary', in that the surfaces have never borne a previous vegetation cover. However, the actual particles of the new soil may well have been part of an old soil elsewhere, thus making the ensuing succession arguably 'secondary'. Vegetated sand dunes can be destroyed by wind and the sand shifted to form new dunes. An estuarine salt marsh lies on soil washed downstream by the river, or eroded from another marsh [86]. Much of the new surface after a landslide, even if not covered with scattered sods from the old vegetation, may contain roots and is hardly a 'virgin' substrate. Thus the division between primary and secondary successions seems as arbitrary as that between successions and fluctuations, though it can still be convenient. What counts is the inherent chemical and physical suitability of the exposed substrate for plant growth. A new sterile surface like a glacial moraine or a quarry floor may be colonized much more rapidly than a bare patch of soil overlying serpentine rock, or on a peat bog.

2.1.2 Climatic processes

Climatic processes initiating vegetation change include: drought [54, 69]; fires caused by lightning [57, 96–98]; wind, causing falling of trees [99–102] or excessive desiccation of evergreens in winter [103, 104]; winter cold, including early autumn and late spring frosts, snow avalanches, snow drifts [105] and glaze storms [104]; and long term climatic change [106].

The effects of drought and winter desiccation may be gradual and cumulative, but other climatic processes causing change in vegetation are frequently abrupt and catastrophic in their effects. The effects of long term climatic change may be too slow to be measurable within a human life span, but they may be inferred from palynological records.

2.1.3 Biotic processes

Biotic processes causing change in vegetation include: the effects of other plants (see Sections 2.4 and 2.5); grazing by herbivores, from insects [107, 108] to mammals [109, 110]; plant disease epidemics [111]; and the numerous effects of man, both direct, like ploughing soil, using herbicides or felling trees, and indirect, like causing fires, eutrophication of water bodies, soil, water and air pollution, or influencing grazing pressures.

The effects of 'natural' biotic influences tend to be gradual (insect and diesease epidemics do not occur overnight), but those of man vary from the barely significant to the abrupt and catastrophic.

A more detailed account of these various environmental influences is given by Egler [29].

2.2 Immigration of species

All parts of the world appear to be subjected to a 'rain' of plant propagules. These are carried by wind, water and animals, and nowadays in more unusual ways also, for example in the packaging around trade goods, or in mud on motor vehicles.

2.2.1 Patterns of dispersal with distance

The seed rain is of course heavier in densely vegetated than in sparsely or non-vegetated areas. Dispersal tends to decline logarithmically with increasing distance from the source. Fig. 2.1 shows the dispersal pattern of wind-dispersed nutlets from a single birch tree. If the nutlet density on the ground is replotted against the logarithm of the distance from the tree, a straight line results. (The aberrant first point is due to the sampling position being at the edge of the canopy which reduced the fall of fruits underneath.)

As propagule dispersal tends to appear as a negative function of the logarithm of the distance from the source, it is not surprising that species colonization, i.e. establishment after dispersal to a given site, also tends to show a similar relationship. An example of this comes from the work of Maguire [113], who set out jars of nutrient-enriched water at varying distances from a pond in Texas. He then recorded the appearance in them

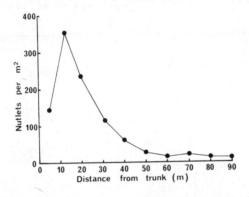

Fig. 2.1 The dispersal pattern of nutlets from a single birch tree (*Betula pendula*). (From Sarvas, 1948 [112], courtesy of the Finnish Forest Research Institute.)

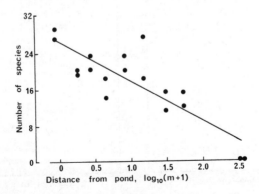

Fig. 2.2 Relationship between the number of microbial species colonizing jars of water after 25 days and the distance of the jars from a pond in Texas. (Data from Maguire, 1963 [113]).

23

of different microbial species. These had presumably been dispersed mainly as air-borne spores originating from the pond. Fig. 2.2 shows that after 25 days the numbers of micro-organisms present clearly declined with increasing distance from the pond.

The size of propagules and their method of dispersal obviously influences the effective distance travelled. Thus Fox [115], studying the natural regeneration of dipterocarp rain forest in Sabah, found that patterns of seedling distribution on the ground were related to fruit characteristics. Seedlings from fruits with large wings tended to be regularly distributed within the forest, those from small-winged fruits were less regularly distributed, while those from wingless fruits showed a markedly clumped distribution.

Dispersal of seed over the ground within the effective dispersal distance from a parent plant tends to be essentially a random process. For example, when Brereton [116] studied colonization by *Puccinellia maritima* and *Salicornia europaea* on coastal mudflats at Foryd Bay, North

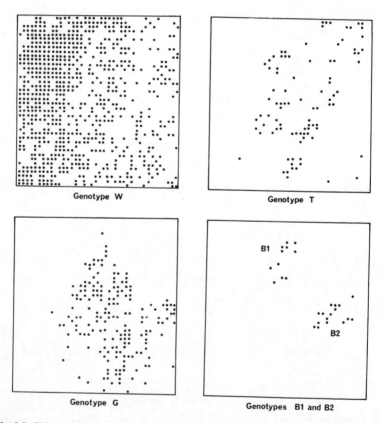

Fig. 2.3 Clonal growth: distribution of five genotypes of *Festuca rubra* within an 84 m² plot in the Pentland Hills, Scotland. (From Harberd, 1961 [114], courtesy of Blackwell Scientific Publications.)

24

Wales, he found that the initial distribution of both was random, though clumps later formed and eventually coalesced. However, seeds may be blown or washed about on the ground after they have first arrived there from the parent plants. Topographic discontinuities then determine where seeds finally come to rest. Thus when the *Calluna*-dominated vegetation of heathlands and moorlands in Britain regenerates after destruction by fire, most seedlings are found in small cracks on the burnt soil surface. On a larger topographic scale, concentrations of *Pinus sylvestris* saplings can sometimes be seen locally along the steep sides of morainic banks and hummocks in the Scottish Highlands. These patterns result from cones of *P. sylvestris* having released seed when there was a glazed snow cover, seed then being blown over the snow surface and lodging where there were abrupt changes in relief.

Vegetative spread by clonal development can be a surprisingly effective method of dispersal. Harberd's work on the grass *Festuca rubra* [114] is a good example. Harberd collected tillers from an 84 m^2 patch on a grassy hill, and by detailed morphological examination, backed up by studies of self- and cross-compatibility, was able to identify seventeen distinct genotypes or clones. Several of these were abundant within the patch (Fig. 2.3), and must therefore have spread several metres at least. When he investigated genotype distribution beyond this small patch, he found one genotype with a spread of at least 64 m, while genotype W (Fig. 2.3) was spread within an area of more than 200 m diameter. He estimated that this latter would have taken between 400 and 1000 or more years to achieve.

2.2.2 Patterns of colonization with time

Species colonization with time tends to change logarithmically with many species colonizing early on, but with species additions gradually decreasing with time. This pattern results from the fact that common species, producing an abundant seed rain, will tend to arrive early and to colonize in the first few years of a succession provided that conditions are suitable. With progressively rarer species however, the probability of a propagule falling on any given patch decreases, so they tend to colonize over a longer period of time.

Examples of colonization with time are shown in Fig. 2.4: (a) represents colonization of seven man-made ponds of different ages lying within a 4 km stretch beside the river Trent in England [117]; (b) represents colonization of a series of sand dunes at Grand Bend, Ontario [118]; (c) represents colonization of Krakatau after the volcanic eruption of 1883, which destroyed half the island and covered the rest with tens of metres of hot lava and pumice, so that the entire flora was almost certainly destroyed [119]. Colonization of both the pond and sand dune series showed a logarithmic relationship with time. That of Krakatau differed in showing a linear relationship, though this probably represents only the early phase of rapid arrival of new species. Interestingly, in neither the pond nor sand dune series do the regression lines approach the origin. Assuming that the ages of the two series were accurately determined, there seems to have been a lag in early col-

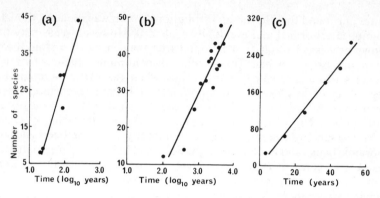

Fig. 2.4 Relationships between time and the numbers of vascular species present in primary successions (a) in man-made ponds in England, (b) on sand dunes in Ontario, and (c) on the volcanic island of Krakatau. (Data from Godwin, 1923 [117], Morrison and Yarranton, 1973 [118] and Docters van Leeuwen, 1936 [119] respectively.)

onization in each case. This could have resulted if a period of physical stabilization was necessary before colonization was possible, or if site modification by pioneer species eventually facilitated the establishment of others.

However, colonization seems more usually to be rapid from the outset. On Krakatau, twenty-six vascular species were found growing only three years after the eruption, and sixty-four after fourteen years. This was in the species-rich tropics, but species colonization can be relatively as prolific in the species-poor subarctic. Thus when the new island of Surtsey was built up by undersea volcanic eruption off the coast of Iceland from 1963, the first vascular species was found growing there only eighteen months after the onset of the eruption. By the end of 1967, the year eruptions ended, seeds and parts of twenty-seven vascular species had been found where they had drifted ashore, while five vascular species and two mosses were found growing [120].

The state of flux noted earlier to be a characteristic of all vegetation is evident as soon as colonization of bare ground begins. Thus Cooper [79] recorded all individual plants present in nine permanent quadrats on newly deglaciated till and outwash surfaces at Glacier Bay, Alaska. After five years, he found that although there were 18% more individuals present, 68% of the new total had established since the beginning of the period, while 62% of those originally present had died during the five years. MacArthur and Wilson [121] have followed up population studies like this with an interesting theoretical study of colonization. This was based on the assumption that species addition curves like those in Fig. 2.4 result from the balance between species immigration and species extinction at any given site. This pictures colonization as the difference between the probabilities of new species establishing and of existing species becoming locally extinct.

2.2.3 Seed production

While vegetative spread can be a very effective method of species dispersal over short distances, i.e. from a few centimetres to a few metres per annum, and indeed may be the main method of regeneration in many grasslands for example, most dispersal over longer distances is by seed. In many species, the production of viable seed fluctuates markedly from year to year, and years with abundant fruiting may be infrequent [122]. Annual variations in seed production are influenced by a variety of climatic factors. Superimposed on these fluctuations are variations in seed losses on the plant due to seed eating insects and birds, and to disease, losses which may be considerable. However, little is known about any of these processes for most species.

The fluctuating and uncertain nature of seed production, from plants which are themselves usually distributed irregularly in space, combined with the random nature of seed dispersal, must result in a seed rain showing almost limitless variations in time and in space in the combinations and relative abundance of the species represented. An unpredictably varying seed rain makes it very much a matter of chance which species arrives first at any given gap in a patch of vegetation, thus getting a head start in the race for establishment and for pre-emption of that space.

2.3 Establishment

Establishment from seed comprises the phases of germination, seedling survival, and subsequent onward growth of the plant. The environmental requirements for these, together with those for seed production and dispersal, form what Grubb has termed the *regeneration niche* of a plant [122]. Establishment is a particularly vulnerable phase of a plant's life cycle. Seeds may be eaten on the plant or after dispersal by a variety of vertebrate and invertebrate herbivores [123]. If seeds survive such predation, they will germinate if they have landed in a *safe site* [124], i.e. a microsite suitable for germination and early seedling survival. Requirements for germination vary greatly between species [122, 125], and thus the environmental specifications of safe sites will also.

Very few seedlings ever survive to maturity. Causes of death include drought, disease, frost heaving, grazing by insects, molluscs and other small animals, failure to grow out of shade, burial under plant litter, faeces, or soil moved by burrowing animals, sun scorch, and mechanical damage by animal hooves. In practice, combinations of these may be responsible. For example, shaded seedlings are more susceptible to disease [126] and grazing [127].

In undisturbed vegetation with a closed canopy, very low seedling densities are generally found [128]. It seems that the establishment of autotrophic plants in vegetation occurs only in gaps, i.e. bare areas formed by the deaths of previous occupants, or where there is incomplete cover. Many of the agents baring substantial areas of ground were mentioned in Section 2.1. Small gaps can be created by plants dying of old age, disease or other causes, or by soil being bared by animal hooves or burrowing animals.

Table 2.1 Mean percentage cover after three years of natural recolonization of different sized patches of ground bared in *Calluna* heathland in north-east Scotland. (From Miles, 1974 [129], courtesy of *Journal of Ecology*.)

	Patch size (cm^2)		
	2500	250	25
Phanerogams regenerated from seed	35	22	7
Phanerogams regenerated vegetatively	20	40	52
Moss	14	40	49
Bare ground	34	8	4

Small gaps generally have a more equable microclimate than large gaps, and for most species will tend to favour germination and early seedling survival. Thus when Miles [129] created three sizes of gaps in *Calluna* heathland, 25, 250 and 2500 cm^2 in area, he found that the mean establishment of self-sown seedlings after one growing season was 56, 34 and 10 plants per m^2 respectively. The most extreme differences were shown by *Calluna*, with 480, 359 and 10 plants per m^2 respectively. However, gaps are evanescent, and tend to become quickly overgrown by vegetative spread of surrounding plants. Thus in the same experiment, survival trends after two years were reversed. For example, survival of first year *Calluna* seedlings for a further two years was 0.5, 4 and 66% respectively in the three gap sizes. Three years after the gaps were created, there were large differences between the three gap sizes in the origin of the regenerated plant cover (Table 2.1). The largest size had almost twice the cover from seed than from vegetative regeneration, whereas in the smallest size there was over seven times more cover from the latter than from seed. This illustrates the point that the optimal conditions for survival and growth of a seedling may change as a plant matures. In this experiment it was clearly disastrous for seed of most species to meet optimal conditions for early establishment.

The size of gap or degree of 'incomplete coverage' necessary for establishment differs markedly between species. Table 2.2 gives some of the results from another experiment in *Calluna* heathland in which seed was sown into untouched vegetation, and into vegetation with the *Calluna* canopy, the ground layer, and the litter layer successively removed. The increase in establishment of particular species with each successive alteration of the vegetation structure indicates that effective gaps or safe sites had been created.

Seedlings of most species cannot grow in heavy shade, so for early survival and subsequent growth the seedling shoot must grow into or above the canopy. Whether or not a seedling can grow out of shade where it is below its compensation point depends on the height and shade profile of the canopy, and on its ability to produce extension growth by retranslocation of organic materials. Salisbury [130] long ago showed that, in Britain, plants of closed vegetation, like woodland, tended to have heavier seeds than plants of open or pioneer vegetation. He suggested that the capacity of autotrophic plants to establish in the face of competition for

Table 2.2 Mean establishment after one growing season from seed sown into 1 m² plots of *Calluna* heathland in north-east Scotland with different layers of the vegetation removed. (From Miles, 1974 [129], courtesy of *Journal of Ecology*.)

	Control	Calluna canopy removed	Calluna canopy and ground layer removed	Canopy, ground layer, and litter removed
Agrostis tenuis	0	0.5	4	24
Deschampsia flexuosa	0	0	2	5
Holcus lanatus	0	1	0.8	23
Hypochoeris radicata	0	2	4	22
Luzula sylvatica	0	0.2	0.5	14
Rumex acetosa	0	0	0.5	27
Sarothamnus scoparius	0.5	4	6	10
Ulex europaeus	0	0	1	6

light was associated with the size of the seed's food reserves, and that heavier seeds thus had an initial advantage. Grime and Jeffrey [131] have since shown that in many species the ability of a seedling to produce extension growth in shade is correlated with the amount of stored reserves in the seed (Fig. 2.5). Further, seedlings with small seed reserves showed a higher death rate than seedlings with larger reserves (Fig. 2.6), generally succumbing to fungus infections.

Baker [132] has since examined the seed weights of nearly 2500 taxa growing in California in relation to the environmental conditions in

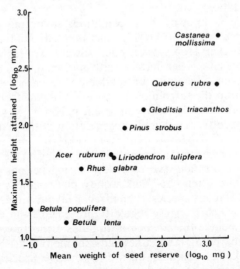

Fig. 2.5 The relationship between log maximum height attained by seedlings of different tree species after 12 weeks in a 55 cm shade stratum, and log mean weight of the seed reserve. (From Grime and Jeffrey, 1965 [131], courtesy of *Journal of Ecology*.)

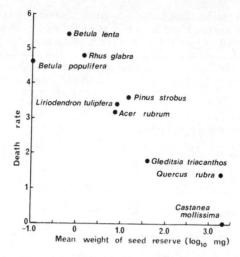

Fig. 2.6 The relationship between death rate (fatalities per container over 12 weeks) of seedlings of different tree species growing in a 55 cm shade stratum, and log mean weight of the seed reserve. (From Grime and Jeffrey, 1965 [131], courtesy of *Journal of Ecology*.)

which they normally grow. He found the correlations with shade conditions less marked than Salisbury had in predominantly mesic England. Instead, there was a marked tendency for seed weights to increase with increasing likelihood of the seedling being exposed to drought after germination. Greater seed reserves presumably facilitate the rapid development of a seedling's root system, leading to better drought avoidance.

Rather little is known about conditions favouring the growth of seedlings to maturity. Grubb [122] has pointed out that the optimal conditions for several forest trees differ between the seedling and sapling stages, while we saw earlier that, in heathland, small gaps favouring early establishment may be quite unsuitable for later growth because of competition from the surrounding vegetation. Apart from competition, grazing and fire often have an important influence, at least on woody plants. In the Scottish Highlands, regeneration of the main native trees, *Betula pendula*, *B. pubescens*, and *Pinus sylvestris*, and the large shrub *Juniperus communis*, is often prevented by browsing on the saplings by red deer and sheep, on *Juniperus* especially during periods of snow lie in winter [109]. Fire often cuts back woody plants, and can prevent the establishment of many species (though it can favour others). Examples of vegetation in which trees are regularly prevented from establishing by periodic fires include the *Calluna* heathlands of northwest Europe, the Californian chaparral, and grasslands in many parts of the world [133, 134, 135]. An example of a fire-sensitive species is *Juniperus occidentalis*. Stands of this in southwestern Idaho were apparently restricted by periodic fires before white settlement to rocky ridges where fires were less intense [136]. With active fire control over the past century,

30

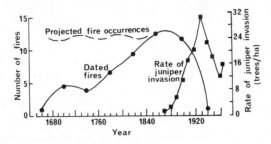

Fig. 2.7 Relation between frequency of fire and rate of invasion by *Juniperus occidentalis* in southwestern Idaho. (After Burkhardt and Tisdale, 1976 [136].)

the juniper has spread into sagebrush-grass vegetation where it was previously excluded by fire (Fig. 2.7).

2.4 Competition

Competition may be defined as the hardships which result to organisms from the proximity of neighbours [137, 137a], or as the consequences when one individual is sufficiently close to another to modify its soil or atmospheric environment and thereby decrease its rate of growth [138]. Such a definition covers two classes of situation. Firstly, where use of a given environmental resource by one individual reduces the resource available to another, as in competition for light or water. (The use of the term competition is sometimes restricted to this meaning [33]). Secondly, where some by-product of an individual's growth modifies the environment of another, thus influencing its growth, e.g. by toxic exudates [139], soil acidification [140, 141] or smothering litter [142]. Such indirect effects of plants upon others should logically include the case where a patch of vegetation containing one or more preferred species, adapted to heavy grazing, attracts a high grazing pressure by a particular large herbivore, with the result that less preferred species surrounding the patch will also be grazed more intensively than elsewhere, possibly causing a gross change in species composition [109]. This happens frequently in upland Britain, where patches of *Agrostis-Festuca* grassland can, with moderate sheep grazing, expand at the expense of the surrounding *Calluna* heath. Another case would be when a particular dominant produces litter or standing material which is especially inflammable, supporting frequent fires which kill less fire-resistant species. Examples of dominants maintained by fire, originally by natural fires caused by lightning and aboriginal man, include the *Pinus ponderosa* forests of the western United States, and *Adenostema fasciculatum* (chamise) and other shrubs of the Californian chaparral [134]. These latter, more indirect influences of particular species on others, are not usually considered as competition (in fact, they have generally been overlooked), but they can be of overwhelming importance.

Because of the many ways in which plants use and modify their environment, competition is a complex phenomenon. All the individuals

31

in any patch of vegetation will be seeking light, water and mineral nutrients, and all will be releasing a variety of organic compounds, some toxic, into the soil, via their leaves, roots, mycorrhizas and rhizosphere organisms, and as products of the decomposition of their litter. Due to technical difficulties in controlling all these factors together under experimental conditions, there is a dearth of knowledge about precise mechanisms of competition in field situations. Competition is thus a much used but poorly understood term.

It can readily be inferred that competition is of critical importance in determining the species composition of vegetation. For example, Byer [143] studied an area where *Pinus banksiana* forest on a sandy soil graded into a *Sphagnum-Chamaedaphne* bog, and Miles [144], an area of heathland where four distinct types of vegetation were found on four different soils. Both workers grew a number of species from seed on bared soil in the different vegetation types, and found that most species established and grew at least as well on soils where they did not occur naturally as on soils where they did. The inference is that interspecific competition was the main factor determining their natural distribution. An indication of the severity of competition was given by Ross and Harper [145], who found that the weight of seedlings of *Dactylis glomerata* grown in different sized patches was proportional to the cube of the patch radius, i.e. of their mean distance from neighbours.

Competition for light seems to be important wherever vegetation completely covers the ground. A simple experiment showing its importance was carried out by Black [146]. Using the same genotype of *Trifolium subterraneum*, he sowed swards of equal density with small seeds (mean weight 4 mg), large seeds (10 mg), and in mixture. Plant weights at harvesting were the same from both small and large seeds when grown in pure stands. But in mixture, the plants from large seeds had become dominant, with a mean dry weight per plant of 0.32 g, whereas plants from small seeds weighed only 0.06 g, and were so heavily shaded that they received only 2% of the incident light. The success of the plants from the large seeds was because from an early date they had a slightly greater area of leaves borne on slightly longer petioles.

From the middle of the last century, it was thought that competition was mainly for light, until Fricke [147] demonstrated by a root trenching experiment in 1904 that root competition from a *Pinus sylvestris* stand was inhibiting the growth of woody and herbaceous species under it. Root competition under different forest canopies has since been demonstrated many times [148, 149, 150, 151, 152], though less is known about the effects of root competition between herbaceous species. Korstian and Coile [152] trenched around plots of ground vegetation to exclude tree roots under six different kinds of woodland in the Duke Forest, North Carolina, and obtained some profound results. For example, Fig. 2.8 shows that under a 31-year-old stand of *Pinus taeda* there was a striking increase in the number of individual plants on the ground, in the number of species present, and in the growth response of tree seedlings. Early studies suggested that effects like this seemed to be due to competition for

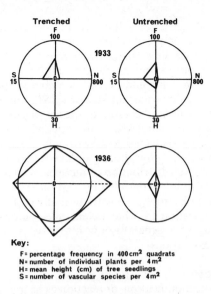

Trenched   Untrenched

1933

1936

Key:

F = percentage frequency in 400 cm² quadrats
N = number of individual plants per 4 m²
H = mean height (cm) of tree seedlings
S = number of vascular species per 4 m²

Fig. 2.8 Changes in the ground vegetation under a 31-year-old *Pinus taeda* stand after eliminating tree root competition by trenching. (After Korstian and Coile, 1938 [152].)

nutrients rather than water (151, 153, 154), a conclusion borne out by a more recent study using fertilizers labelled with ^{15}N and ^{32}P [155].

2.4.1 Allelopathy

Many plants produce toxic chemicals which inhibit the growth of other plants [134, 156], a process called *allelopathy* [157]. Such chemicals may be autotoxic, or may affect other species, often in a highly selective manner. Allelopathic chemicals are just part of a much greater array of secondary plant compounds that are thought to influence other organisms [158, 159], including those which seem to act as a chemical defense against micro-organisms and herbivores [160, 161, 162].

Most plants contain phytotoxins, and Whittaker [158] has contended that allelopathy is 'a widespread and normal, although mostly inconspicuous, phenomenon of natural communities'. However, the existence of a toxin in a particular plant does not mean that it is released into the environment of another plant in concentrations sufficient to affect it in any way. Further, the technical difficulties of proving an allelopathic mechanism to exist in a particular field situation are immense. It is thus not surprising that there are few well established cases of allelopathy occurring in the field rather than merely inferred from laboratory and glasshouse studies. Some cases seem fairly incontrovertible, however; for example, the inhibition of herbaceous species by many shrubs of the Californian chaparral [163]. Allelopathy also seems to account for the well-known inhibition of growth of certain trees, most notably *Picea abies*, on *Calluna* heath (termed 'checking' by foresters), with the *Calluna*

33

roots producing a substance inhibitory to the mycorrhizal fungi of *Picea* [164, 165]. An example of self-exclusion by allelopathy is the killing of *Grevillea robusta* seedlings under the parent trees [166].

Whittaker's contention that allelopathy may be almost everywhere involved, *inter alia*, in controlling vegetation composition, is supported by some recent work by Newman and Rovira [167]. This showed that root exudates of some common grassland species occurring naturally as isolated individuals were somewhat autotoxic, whereas the reverse was true of certain species that tend to be strongly gregarious. These root exudates seem to affect phosphorus uptake [168], and may help to account for the observed distribution of the species concerned. However, effects at this level cannot be detected by field observations, unlike those noted in the previous examples, so without many more instances it must remain an open question to what extent allelopathy is a ubiquitous influence operating at a generally low intensity, or a fairly dramatic influence in a few situations.

2.5 Site modification

Clements [2] saw site modification, or reaction as he termed it, as the main driving force of succession, but he seems to have greatly overemphasized its importance (see Section 1.3.2). For example, he believed that successions on bare rock could begin with lichens, and that plants themselves could weather rock into soil. As early as 1912, Bassalik showed that bacteria could accelerate the natural weathering of rock [169, 170], and similar effects of bacteria, fungi and lichens have been shown since [171, 172, 173, 174]. However, despite this, it now seems that lichens growing on bare rock form effectively a permanent stage [175, 176, 177]. Further vegetation development cannot progress until soil is present, and this is formed chiefly from the weathering of rock by physiographic processes. Again, Clements believed that succession in open water (hydroseres) resulted mainly from the deposition and accumulation of plant debris, whereas most seem to be strongly influenced by the deposition of silt brought from elsewhere [83], and successional change may be rapid only where such deposition occurs [178].

The environment of any site is influenced by vegetation, however, and influenced differentially by different kinds of vegetation. As soon as plants colonizing bare ground form a vegetation cover, the microclimate is modified greatly [179, 180]. Many plants tend only to grow in the shelter of a vegetation cover, the numerous bryophytes that need a moist atmosphere, for example [181]. Others may do so when conditions are otherwise unfavourable. Thus, in the Jezira desert of central Iraq, perennial plants tend to trap small mounds of wind-borne soil which then support a rich growth of annuals [182]. Grazing often prevents re-generation of woody species. In the Scottish Highlands, individual trees, especially of *Sorbus aucuparia*, often establish within dense bushes of *Juniperus communis*, where they are protected from grazing, and eventually the saplings grow up through the canopy. Similar examples have

34

been recorded elsewhere, for instance, in the heavily grazed Middle East [182, 183]. The type of vegetation also greatly influences the development of soil profiles, and the physical and chemical characteristics of the surface horizons [184, 185]. In Chapters 5 and 6 we will see examples in which soil changes may facilitate the establishment of new species, and thus continue succession. In general, however, site modification seems to have rather little influence on succession.

2.6 Stabilization

Clements [2, 42] argued that succession led to the formation and stabilization of a climax type of vegetation. Stability for Clements meant persistence in time: '. . . climaxes are characterized by a high degree of stability when reckoned in thousands or even millions of years' [42]. However, it was seen in Section 1.3.2 that vegetation stability is often illusory. Climax vegetation, if such be deemed to exist, is much more variable and dynamic that Clements seems to have acknowledged.

The subject of stability in vegetation and ecosystems is complex, and has bred an extensive literature [186, 187, 188], though this contains many untested theoretical models but rather little empirical data [189]. Stability can have many interpretations. As well as persistence in time, Orians [190] has listed six additional meanings; others are possible [191, 192]. The literature on the relationship of stability to other ecosystem attributes, e.g. diversity, is ambiguous. For example, the traditional belief among ecologists that complex kinds of vegetation or ecosystems are more stable than simple ones, appears, on cursory examination, to be rejected by some workers [188, 193, 194] but still accepted by others [189, 195]. However, the definitions of stability used can be quite different, and thus not mutually exclusive. For example, McNaughton [189] defines it as stability of community functions, e.g. energy flow, and has some experimental evidence that such stability increases with time during succession. In contrast, Horn [194] defines it as the speed of return to equilibrium of a community after disturbance, and notes that this must decrease during succession. 'Disturb early succession and it becomes early succession. Disturb a climax community and it becomes an early successional stage that takes a long time to return to climax.' An understanding of ecosystem stability is of vital importance to man in the fields of food production and forestry, and it is clear that there is still much to learn.

3 Fluctuations

3.1 Definitions of vegetation change

It will now be helpful to elaborate further on different kinds of change in vegetation. It was noted in Section 1.3 that the literature on the terminology of vegetation change is confused. This is partly because vegetation is infinitely variable in space and time, so that there are no hard and fast dividing lines between any different kinds of change that may be recognized, and partly because our perception of change varies with the scale at which vegetation is viewed.

All changes result from the continuous flux in populations of individual species that make up the vegetation. When vegetation keeps the same overall composition in terms of the combinations and proportions of the species present, the replacement processes may be thought of as *regeneration*. *Fluctuations* were earlier defined as reversible changes about a notional mean, and *successions* as directional changes away from an initial state (see Section 1.3). The distinction was also made between *primary* and *secondary* successions (Section 2.1.1), though it was stressed that this also was arbitrary, and that in reality the difference was often blurred.

When isolated individuals in vegetation die and are replaced, the regeneration process may be barely noticeable. However, sometimes large individuals, or groups of individuals, in areas varying from a few square decimetres to a few hectares, die and are replaced, giving rise to a mosaic of different phases of regeneration. Further, the previous occupants of these gaps may be replaced only after another species, or a species of contrasting life form, has occupied the gap for a term. The changes are cyclic processes, but have sometimes been wrongly termed cyclic successions, when in fact they are just a more extreme form of 'community' regeneration. Even more confusing, however, are situations where natural regeneration processes are indistinguishable from secondary successions. These just stress the artificial and arbitrary nature of our terms. For example, the forests throughout much of North America were historically, and still are, variously and periodically burnt by natural and man-made fires [57, 96, 134], felled by wind [100, 196] and destroyed by insect attack [107]. When any given patch is destroyed a series of species population changes will (unless there is further disruption) eventually reproduce something like the original vegetation, and as such is secondary succession. However, if viewed over a whole forested region, the process becomes merely regeneration, and the forest a mosaic or cyclic series of different phases of regeneration. Which term is preferred depends solely on the scale adopted, though here we are concerned with the patch rather than the region, and will thus consider this as secondary succession.

3.2 Phenological changes
Most vegetation is characterized by marked differences in aspection, that is, in the seasonal development or phenology of the component species. Examples of vegetation in which such changes have been well documented include the American prairie grasslands [31, 197] and tropical rain forest [198, 199]. Aspectional changes can greatly alter the appearance, and sometimes the apparent composition of vegetation. Fig. 3.1 shows the successive periods of leafing out, flowering and fruiting of a group of species in a temperate grassland. Botanical analysis of such grasslands at different times of year can show considerable variations in terms of percentage cover of different species, or their percentage dry weight contribution to the total standing crop. This should be remembered in any exercise aimed at monitoring long term changes in vegetation.

3.3 Changes with fluctuations in environment
Fluctuations occur in the biotic and climatic components of the environment. These affect plants at all stages of their life cycles. Fluctuations occur in the populations of invertebrate and vertebrate herbivores, and of microbial pathogens, which influence the abundance and relative performance of different species [107, 111, 122]. In general, however, these are probably less important in causing fluctuations in vegetation than climatic fluctuations, which have a particularly marked influence on the balance of competition between species in the vegetative phase.

Temporal fluctuations in climate occur on varying time scales. Seasonal changes within the year cause the phenological changes noted in the

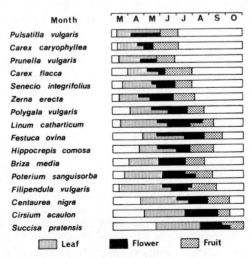

Fig. 3.1 Phenology of sixteen species in chalk grassland in southern England. (From Wells, 1971 [200], courtesy of Blackwell Scientific Publications.)

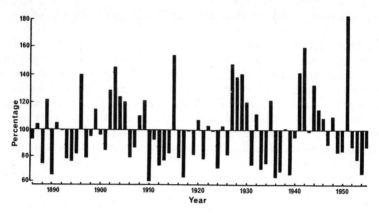

Fig. 3.2 Departures from mean annual precipitation at Salina, Kansas, 1886–1955. (From Jenks 1956 [202], courtesy of the University of Kansas.

previous section. Very long term fluctuations caused the glacial cycles, which had a dramatic effect on vegetation (see Chapter 5), and are presumably still slowly occurring. Changes occur over centuries. A climatic deterioration in Europe during 1100–1300 contracted the limits of cultivation of crops like wheat and grapes, lowered the natural tree line on mountains, and probably also influenced the distribution of many other species [201]. However, such fluctuations over centuries and millenia are probably best considered as a particular kind of succession. Unlike very short-term fluctuations they are probably never fully reversible, because of concurrent changes caused by physiographic and pedological development, and by immigration, extinction and evolution of taxa. Vegetation fluctuation as a term is best restricted to changes occurring from year to year, or over periods of a few years only.

Many climatic factors causing change in vegetation were listed in Section 2.1.2. These may variously cause fluctuations or successions. Their effects may be dramatic. In the boreal taiga, periods of especially extreme cold in winter and early spring can kill large numbers of normally frost-resistant trees [203]. Effects due to fluctuations in rainfall are probably most important on a world basis, however. Annual rainfall values often show deviations of 25–30% or more from long-term means [205, 206], especially in arid and semi-arid areas where the effects of below average rainfall on vegetation are most profound.

The effects of drought on vegetation have been best documented for the Prairie and Great Plains grasslands of North America [54, 55, 197, 207]. These are characterized by widely varying annual rainfall [208]. Fig. 3.2 shows fluctuations at Salina, Kansas. It shows the tendency for alternating periods of mainly wet and mainly dry years, though the pattern is interrupted, for example by the wet years of 1896, 1915 and 1951 within otherwise dry periods. The composition of the prairie grasslands changes dramatically as a result of these recurring droughts. For example, Fig. 3.3 shows the changes from 1932 to 1961 in the two dominant species in a

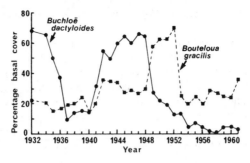

Fig. 3.3 Changes in percentage basal cover of the two dominant grasses in short grass prairie, Kansas, 1932–1961. (After Albertson and Tomanek, 1965 [204].)

patch of short grass prairie. During the 1930s drought, cover of *Buchloë dactyloides* (buffalo grass) was greatly reduced. It quickly recovered after the drought, but again decreased during the 1950s drought. Cover of the codominant *Bouteloua gracilis* (blue grama) tended to follow a reverse trend, because, as a rhizomatous species, it is more resistant to drought than the stoloniferous *Buchloë dactyloides*, thus gaining a competitive advantage when cover of the latter is reduced [207, 209].

Pronounced fluctuations also occur in the composition of temperate grasslands [210, 211, 212]. Table 3.1 shows changes in the annual productivities of the most abundant species in a flood meadow. Many of the species showed remarkable year-to-year fluctuations, and longer term trends, in their relative contributions to the sward. Such variations are probably caused mainly by the differential effects of yearly differences in weather on the competitive ability of species. For example, in such meadows *Agropyron repens* and *Zerna inermis* tend to predominate in dry

Table 3.1 Changes in annual productivities (g per $8\,m^2$) of the most abundant species in a flood meadow, Oka River, Russian S.S.R. (From Rabotnov, 1966 [211], courtesy of W. Junk.)

Year	1954	1955	1956	1957	1958	1959	1960	1961	1962	1963
Agropyron repens	137	162	70	123	131	141	45	49	60	36
Agrostis stolonifera	286	355	24	7	75	62	28	35	34	160
Festuca pratensis	162	210	90	180	210	226	85	217	219	433
Phleum pratense	262	230	39	47	140	135	25	64	32	16
Poa pratensis	147	151	75	126	365	369	112	392	307	223
Zerna inermis	174	166	113	140	132	141	78	85	47	40
Lathyrus pratensis	81	149	130	237	305	180	115	119	108	132
Trifolium hybridum	24	7	78	152	74	5	5	4	1	1
T. pratense	251	31	251	313	56	19	9	50	50	52
Vicia cracca	59	56	105	177	114	85	57	51	86	151
Achillea millefolia	99	165	146	215	225	79	53	123	197	120
Campanula glomerata	3	19	18	71	144	133	109	306	126	97
Cirsium setosum	220	536	261	455	275	130	84	65	84	96
Seseli libanotis	56	51	56	132	140	87	88	109	56	57
Silene vulgaris	4	26	9	44	49	71	68	88	81	22
Total production	1967	2314	1465	2419	2435	1863	957	1757	1488	1636

39

years, while in wet years *Alopecurus pratensis* becomes more abundant [35].

Although short term fluctuations in weather can have pronounced effects on herbaceous swards, little is known about their possible effects on woody vegetation, e.g. forest, where the life span of individuals may be 200–300 years or more. However, there is some evidence from a transition zone (or ecotone) between deciduous and coniferous forest in Minnesota that such fluctuations may influence establishment processes, with the result that at times deciduous species may colonize gaps in the coniferous forest, and at other times the reverse occurs [213].

4 Regeneration and cyclic changes

In Section 3.1 we saw that when gaps in vegetation were large enough for regeneration to be very obvious, the changes often appeared cyclic. In this chapter we will discuss a few examples. Regeneration in forest will, with one exception, be dealt with later in Chapter 6.

Cyclic changes are most readily recognized in vegetation dominated by one species. Fig. 4.1 gives a generalized example. Although a gap is usually colonized for a while by species other than the previous occupant, even if this just means an increase in a ground layer of bryophytes or lichens already present, the dominant species may begin to recolonize directly, from seed or vegetative regrowth. Cyclic changes are very obvious in many exposed alpine and arctic areas, where a well marked and dynamic vegetation patterning is created by soil erosion by needle ice and wind, or by soil creep or solifluction [214, 215, 216, 217]. For example, on Mt. Kosciusko, in Australia, cyclic changes occur on plateaux where the vegetation largely consists of isolated clumps of the dwarf shrub *Epacris petrophila* in stonefields. Severe, unidirectional winds simultaneously erode the fine topsoil from the windward edges of clumps and redeposit some of it on the leeward sides. Rooting stems on the windward sides are killed, but plants usually continue to grow, and root by layering in the redeposited soil, on the leeward sides [216].

The dynamics of vegetation of even simple composition are not always so obvious, however. Fig. 4.2 is a simplified map of *Calluna vulgaris* bushes and *Eriophorum vaginatum* (cotton grass) tussocks in an $8\,m^2$ patch of heath on wet peat, where *Calluna* and *E. vaginatum* formed about 75% of the cover. Regeneration appeared to be entirely by vegetative means. Dissection revealed the positions of trailing stems of the former species, and of dead tussock remains of the latter. The dead tussocks indicate the origin of present day tussocks. They also seem to permit vegetative regeneration of the *Calluna* by acting as points above the water table in which the trailing stems can root adventitiously. The trailing

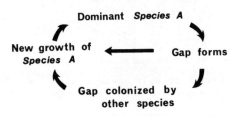

Fig. 4.1 Generalized example of cyclic change in vegetation

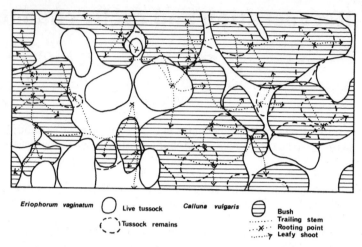

Eriophorum vaginatum ◯ Live tussock   Calluna vulgaris ⊖ Bush
(⌇) Tussock remains   .... Trailing stem
                                            ..x.. Rooting point
                                            ....⌐ Leafy shoot

Fig. 4.2 A simplified map of *Calluna* bushes and *Eriophorum vaginatum* tussocks and remains in a 4 × 2 m plot of wet heath in north-east Scotland. (From Keatinge, 1975 [218], courtesy of *Journal of Ecology*.)

stems further show that apparent bushes of *Calluna* often have more than one contributing source.

Another example of regeneration dynamics which could only have been elucidated through soil excavation is that found by Kershaw in pure stands of *Calamagrostis neglecta* in Iceland [219]. Fig. 4.3 shows a small excavated patch. The notable feature is the very marked clumping of tillers (aerial shoots). Rhizomes may run for a distance equal to several clump diameters before sending up a tiller closely adjacent to existing tillers. Existing clumps are thus continued and enlarged, and may contain

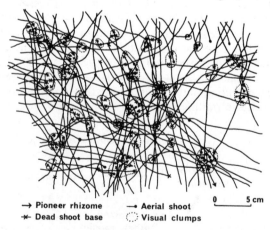

→ Pioneer rhizome   —• Aerial shoot   0   5 cm
✶ Dead shoot base   ⟨⟩ Visual clumps

Fig. 4.3 Distribution of rhizomes and aerial shoots in a patch of *Calamagrostis neglecta* in southern Iceland. (From Kershaw, 1962 [219], courtesy of *Journal of Ecology*.)

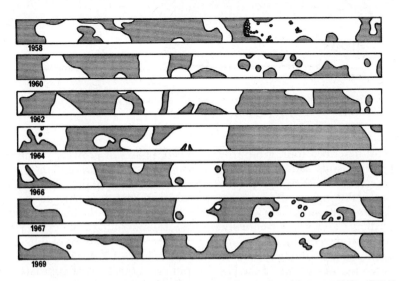

Fig. 4.4 The cycle of change in cover of *Festuca ovina* (hatched area) between 1958 and 1969 in a 10 × 160 cm quadrat within a plot of grassland, fenced against rabbits, in the English Breckland. (From Watt, 1971 [221], courtesy of Blackwell Scientific Publications.)

tillers from several different rhizome systems. The reasons for the formation and continuation of these discrete clumps of tillers in otherwise bare, and apparently uniform soil, are simply not known. This would clearly provide an excellent opportunity for experimentation—no vegetation can be simpler than one with only a single species!

All the examples quoted so far in this chapter have been inferred from spatial sequences observed in the field at a single point in time. Few studies have involved repeated observations through time, notable exceptions being those of A. S. Watt in the English Breckland. For example, in 1935 and 1936 Watt established a number of rabbit-proof enclosures in grasslands on acid soils for long term study. In one of these, the grass *Festuca ovina* had become dominant after eleven years, and eventually eliminated all other species ([220, 221]; and see Chapter 7). Fig. 4.4 shows the annual charts of one small patch, beginning in 1958 when *F. ovina* was already dominant. They show the cycle of change, over about ten years, of the grass population. For example, on the right hand side of the charts there is a scattering of small, young plants in 1958. These had formed complete cover by 1964, but by 1967 had largely died, leaving again a scattering of small plants, some of them young and some shoots surviving from old plants.

Possibly the best known example of cyclic change is that of the dwarf shrub *Calluna vulgaris* [133, 222, 223]. If a *Calluna* stand stays unburnt for long enough, is not heavily grazed, and is not invaded by trees, plants begin to die naturally of old age, and the stand acquires an uneven-aged structure. (Vegetative regeneration by layering, as shown in Fig. 4.2, seems to be largely restricted to wet, peaty areas.) From such a stand,

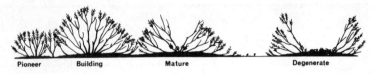

| Pioneer | Building | Mature | Degenerate |

Fig. 4.5 Diagrammatic profile view of *Calluna* bushes showing growth phases. (From Watt, 1955 [222], courtesy of *Journal of Ecology*.)

Watt described four, arbitrarily delimited, phases of growth: young or *pioneer* plants, actively growing or *building* plants, *mature* plants, and senescent or *degenerate* plants (see Fig. 4.5). Watt stated that in the degenerate phase, the branches spread apart and gradually die back, forming a progressively widening gap in the middle of the bush in which seedling *Calluna* may eventually establish, thus initiating a new cycle. This description was obtained by interpreting the spatial pattern of different aged plants as though it represented the phases of a dynamic process. Attention was drawn to the possible pitfalls in this kind of approach in Section 1.3.4. However, better evidence for the cycle has now been produced by following changes over three years in permanent quadrats in the different growth phases [223]. Fig. 4.6 shows the changes in two of

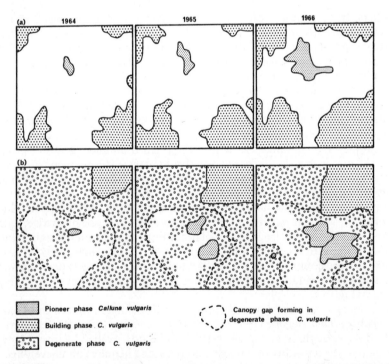

	Pioneer phase *Calluna vulgaris*		Canopy gap forming in degenerate phase *C. vulgaris*
	Building phase *C. vulgaris*	
	Degenerate phase *C. vulgaris*	

Fig. 4.6 Maps of *Calluna* within two 1 m² quadrats on heath in north-east Scotland, recorded from 1964 to 1966. (After Barclay-Estrup and Gimingham, 1969 [223].)

these quadrats. Quadrat (a) shows a small patch of pioneer *Calluna* expanding centrifugally in the middle of an old gap, and also peripheral, building phase plants, expanding centripetally. Quadrat (b) shows degenerate *Calluna* with an expanding canopy gap, together with pioneer and building phase plants as in (a). Changes in the other species present are shown in Fig. 4.7. The important point is that their changes are imposed by those of the dominant species as it goes through its growth cycle.

In the first example in this chapter, the regeneration pattern was imposed by the climate, not the natural growth cycle of the species itself as in later examples. A further, impressive instance of a dynamic, climatically-controlled patterning, is that of the 'wave-regenerated' *Abies balsamea* forests in parts of the north-eastern United States [104]. Numerous, large, crescent-shaped gaps occur in these forests at right

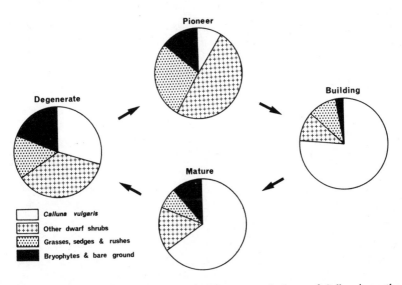

Fig. 4.7 Changes in percentage cover during different growth phases of *Calluna* in north-east Scotland. (Data from Barclay-Estrup and Gimingham, 1969 [223].)

Fig. 4.8 Diagrammatic cross-section through a 'regeneration wave' in an *Abies balsamea* forest in the north-eastern United States. (From Sprugel, 1976 [104], courtesy of *Journal of Ecology*.)

45

angles to the prevailing wind. In profile across these gaps the forest assumes the appearance of successive waves of trees, in which trees continually die off at the front edge of a 'wave' and are regenerated from the seed behind (Fig. 4.8). This patterning apparently takes place because when a gap occurs, trees on its windward edge die from a combination of loss of branches and needles due to heavy ice accumulations in winter, death of needles from winter desiccation, and probably also reduced growth due to greater cooling of needles in summer with the canopy disrupted.

5 Primary successions

Successions have traditionally been termed either primary or secondary, though it was noted earlier (Section 2.1.1) that the distinction was arbitrary, as certain successions could fit either category. The former occur on new, previously unvegetated bare areas, the latter where a previous vegetation cover has been disturbed or disrupted in some way, as by ploughing, fire or changing grazing pressures. It was also noted earlier (Sections 1.3.2 and 2.5) that Clements [2] greatly overemphasized the importance of site modification by plants as a driving force of succession (termed *autogenic* succession, as opposed to successions caused mainly by changes in extrinsic factors, like climate, or the rate of silt input into lakes, termed *allogenic* succession [224]). For example, Egler [50] stressed that vegetation changes during secondary successions on abandoned fields could largely be accounted for by differential growth, reproduction and survival of the species present initially, or colonizing shortly after abandonment (see Section 1.3.2). There seem to be no data that refute this [225]. There is very little evidence about primary successions from direct observations in time, in contrast with secondary successions. Nevertheless, it will be seen in this chapter that in general allogenic factors are probably more important in causing primary successions also, than are autogenic factors.

The profound fluctuations in climate during the last two million years that caused the Pleistocene glacial cycles caused equally profound changes in the world's vegetation [18, 70]. These changes are still presumed to be in progress. An example is the spread of *Picea abies* across Finland, Sweden and Norway after the last glaciation, following changes in the regional climate. This was present in eastern Finland around 3000 B.C., but does not seem to have reached western Norway till about A.D. 1000 and is still absent from parts of that country [227, 228].

Fig. 5.1 is an example of evidence for interglacial vegetation change. It shows variations in the pollen rain of many species, as preserved in lake sediments, at a site in England during the Hoxnian interglacial period, covering a time span of around 40 000 years. These variations reflect changes in vegetation. The site was covered by an ice sheet during part of the previous glacial period, so recolonization was thus, by definition, primary succession. The pioneer vegetation of the late glacial is not represented in the sequence—presumably the future lake was still an area of trough ice then. However, the figure shows that the vegetation changed fairly continuously through the interglacial into the next glacial period. It is noteworthy that there seem to have been no periods when the vegetation could be thought of as stable or 'climax', and no point at which

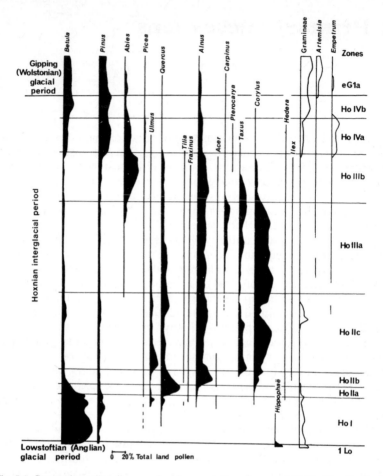

Fig. 5.1 Composite pollen diagram showing vegetation changes at Marks Tey, England, during the Hoxnian interglacial period. (From Turner, 1970 [226], courtesy of the Royal Society.)

the initial primary succession can be seen to have ended. This again highlights the arbitrary and unsatisfactory traditional terminology of vegetation change. Succession was originally defined by Clements [2] as the development of a climax. If climax vegetation is non-existent, but the term succession is retained, the latter becomes meaningless, as all the vegetation changes of the Pleistocene must be 'successional', and all existing vegetation must also be 'successional'. However, as the traditional terminology is still the current terminology, it is retained here. Suggestions for rationalization are made in Chapter 8.

Most of the changes shown in Fig. 5.1 presumably reflect changing climate. Palaeolithic man was present, but is considered unlikely to have had any marked influence on the vegetation. It has been suggested however, that the characteristic increase in *Betula* and *Pinus* towards the end of this and other interglacial periods in north-west Europe may have been caused as much by soil changes favouring these species brought about by the prior dominance of species like *Picea abies* and *Abies alba*, as by climatic change [229]. There is certainly good evidence that the growth of many conifers is associated with marked soil acidification, mor humus formation and accelerated podzolization compared with conditions under most broadleaved species [141, 230, 231]. However, during the last complete interglacial period in Britain, *P. abies* seems to have been uncommon, and *A. alba* absent [18]. Yet, the natural trend in soil development during an interglacial is for progressive leaching with time [232, 233], so the late interglacial vegetation changes were probably caused by a combination of both climatic change and natural soil development, with the latter accelerated when species promoting soil leaching and acidification were present.

5.1 Successions on submerged and waterlogged soils

Most information about vegetation successions on submerged and waterlogged soils (hydroseres) has been obtained from examination of plant remains preserved in stratigraphic sequence in lake sediments or as peat. Using these sources, Walker has made a comprehensive analysis of published records of post-glacial (i.e. Flandrian) hydroseres in Britain that began in confined bodies of freshwater in which allogenic influences,

Table 5.1 Frequencies of transitions between vegetation 'stages' free from obvious allogenic influences derived from 20 pollen diagrams. (From Walker, 1970 [83], courtesy of Cambridge University Press.)

	Succeeding vegetation												
* 'Stages'	1	2	3	4	5	6	7	8	9	10	11	12	Total
1	.	.	.	.	.	.	.	.	.	.	.	.	0
2	.	.	2	4	1	.	3	.	.	.	1	.	11
3	.	4	.	3	3	.	.	.	.	.	.	.	10
4	.	4	1	.	8	.	1	.	.	.	1	.	15
5	.	.	.	.	.	.	6	3	2	.	4	.	15
6	.	.	.	.	.	.	.	.	.	.	.	.	0
7	.	1	.	1	.	.	.	.	6	.	3	.	11
8	.	.	.	.	1	.	.	.	.	.	2	.	3
9	.	.	.	.	.	.	.	.	.	.	6	.	6
10	.	.	.	.	.	.	.	.	.	.	.	.	0
11	.	.	.	.	.	.	.	.	.	.	.	.	0
12	.	.	.	.	.	.	.	.	.	.	.	.	0
Total	0	9	3	8	13	0	10	3	8	0	17	0	71

Left axis label: Antecedent Vegetation

* 1, biologically unproductive open water; 2, open water with micro-organisms; 3, open water with submerged macrophytes; 4, open water with floating-leaved macrophytes; 5, reedswamp; 6, tussock sedge swamp; 7, fen (herbs on organic soil); 8, swamp carr (trees on unstable peat); 9, fen carr (trees on stable peat); 10, aquatic *Sphagnum* spp.; 11, *Sphagnum* bog; 12, marsh (herbs on mineral soil).

such as silting, appeared to be minimal [83]. For convenience, he arbitrarily recognized 12 vegetation stages, though noting that they usually succeeded one another only gradually with time. Table 5.1 gives one of Walker's data sets. This shows an unforeseen variety in the sequence of hydroseral transitions, including the fact that 17% of the transitions were in the reverse direction to that indicated by the great majority of the records. These and other data suggested that if there was an overall trend in these autogenic successions, it was towards *Sphagnum* bog, and not, as had previously been thought, towards scrub and woodland.

The surfaces of *Sphagnum* bogs may become raised beyond the influence of ground water where there is sufficient rainfall to maintain waterlogged conditions suitable for continued peat accumulation. Such 'raised bogs' often have a surface with small pools and hummocks. Earlier workers interpreted these spatial patterns as representing different phases of a system of cyclic change, with pools being colonized, filling in with peat and growing into hummocks, and hummocks eventually degenerating and being eroded to form new pools [82, 234, 235]. However, several recent studies of the peat stratigraphy of raised bogs with such surface hummocks and water-filled hollows have found no evidence for these postulated cyclic changes, but rather that such patterning tends to be persistent over long periods of time, in one case for about 5000 years [236, 237, 238]. This highlights the dangers of incautious interpretation of existing spatial variation in vegetation as representing different phases of a temporal sequence (see Section 1.3.4).

A remarkable example of a persistent pool in peatland is that of the 65 ha Lake Myrtle in the Minnesotan Lake Agassiz peatlands. Heinselman has shown that this has persisted for over 11 000 years, has risen at least 3.6 m with growth of the surrounding peat, and shows no signs of marginal invasion [239]. Heinselman has also shown how the main vegetation changes in this area appear to have arisen from climatic change and as consequences of continued peat development altering surface topography and drainage patterns. A dramatic example of the latter occurred about 3100 years ago when a stream eroded headward into peatland and began to drain much of it in a second direction. This reduced irrigation by ground water, probably causing mineral depletion, and initiated development of the present raised *Sphagnum* bog.

Continuing with peatlands, Drury [95] has described what seems to be a true cyclic succession, driven by physiographic processes, on alluvial areas in Alaska where there exists a mosaic of quaking bogs and *Picea mariana* forest with a thick ground layer of mosses, especially of *Sphagnum* spp. As it develops, this moss layer progressively insulates the soil, eventually causing permanently frozen ground (permafrost). This in turn impedes drainage which further promotes *Sphagnum* growth so that eventually the trees may die—a process termed 'swamping' or 'paludification'. Bogs form by swamping, by colonization of pools, and also by active marginal thawing of existing bogs into frozen ground, which causes ground slumping, swamping, and death of the *P. mariana* trees. With peat

Table 5.2 Changes in percentage cover along 3.2 km at the landward edge of a mature *Spartina* marsh, Bridgwater Bay, south-west England. (After Ranwell, 1964 [240].)

Year	1950	1958	1959	1961	1962
Spartina anglica	100	80	70	54	51
Scirpus maritimus	0	18	25	36	38
Phragmites communis	0	2	4	7	8
Typha latifolia	0	0.6	0.9	3	3

development, forest eventually re-establishes on old bogs and the cycle continues.

Walker's study of autogenic hydroseres [83] also showed large variations in their duration. The succession from open water to *Sphagnum* bog apparently varied from less than 500 years when aquatic *Sphagnum* spp. invaded open water, to 5000 years when the succession passed through a fen stage. However, succession may be rapid when silt is carried into a system and deposited. For example, Ranwell [90] found that up to 10 cm of silt was deposited per year on an estuarine *Spartina* marsh in south-west England. Succession from a pure *Spartina* stand began when the marsh was about 22 years old, but then proceeded rapidly. Table 5.2 shows that within 12 years half of the *Spartina* at the landward edge of the marsh had been replaced by other species.

Succession in saltmarsh is not always progressive, however—silt may also be removed. Thus Johnson and Raup [86] described a *Spartina alterniflora* marsh in a Massachusetts estuary growing on an island of *Spartina* peat. The peat gave no indication that there had been a succession of saltmarsh species. Instead, peat has been continuously eroded from one side of the marsh by tidal scour, and, with some redeposition and growth of new peat on the other side, the result is that the marsh appears to have moved some 260 m eastwards in about 600 years.

5.2 Succession behind retreating glaciers
An early but classic study of vegetation development on newly exposed glacial debris was made by Cooper [79, 80, 81] by comparative observations of sites bared at different times, and also by monitoring changes over a period of 19 years in permanent quadrats in vegetation in an early phase of development. He noted that there tended to be periods of relative dominance by three different groups of species: firstly by herbaceous and mat-forming species, secondly by thickets of *Alnus crispa*, *Salix* spp. and *Populus trichocarpa*, and finally by conifers, mainly *Picea sitchensis* at first, but with *Tsuga heterophylla* and *T. mertensiana* gradually appearing with time. On less well drained sites, *Sphagnum* growth and subsequent swamping may occur later, leading to widespread bog (muskeg) with only scattered trees [241]. Table 5.3 summarizes the main results from the permanent quadrats. During the period 1916–1935 there was a steady increase in the density of herbs and in size and cover of prostrate and mat-

Table 5.3 Changes in species abundance in eight $1\,m^2$ permanent quadrats on recently exposed glacial debris at Glacier Bay, Alaska. (From Cooper, 1939 [81].)

	Year			
	1916	1921	1929	1935
Number of individuals:				
Arctous alpina	0	1	1	0
Polytrichum sp.	0	2	1	0
Poa alpina	1	16	5	0
Epilobium latifolium	110	10	1	0
Habenaria hyperborea	2	2	2	3
Salix spp.	139	140	71	110
Equisetum variegatum	150	161	175	183
Carex spp.	27	43	51	91
Euphrasia mollis	0	17	8	102
Cladonia sp.	0	0	0	13
Pyrola secunda	0	0	0	9
Marchantia polymorpha	0	0	0	1
Aggregate branch length (m):				
Salix spp.	6.5	15.7	21.1	18.4
Cover (dm^2):				
Dryas drummondii	27	120	249	355
Rhacomitrium spp.	40	52	71	127
Stereocaulon tomentosum	0.5	1.4	22	32

forming species. Six new species appeared, though two did not persist, and two of those present in 1916 had also disappeared by 1935. One feature of the results was the slowness with which the glacier had retreated: around 100 km in about 200 years. The large gap at any given time between mature vegetation and newly bared ground will have resulted in a very sparse seed rain at the latter. Thus the establishment of young forest at Glacier Bay took 50 to 100 years as opposed to only 5 to 10 years along the margins of the retreating Juneau ice field about 80 km to the east, where dispersal distances were only a few hundred metres [241].

Despite the appearance of apparent 'stages' during the succession, Cooper stressed that individuals of all stages, including the forest trees, appeared from the outset. Other studies have shown that trees play a pioneering role in primary successions on glacial moraines elsewhere [242, 243], and on volcanic mudflows [87] and probably lava [119] also. It seems clear that, apart from a brief period for physical stabilization of the substrate, no soil development by an earlier vegetation phase is needed for their establishment, in contrast to what Clements [2] thought.

The changing appearance of the developing vegetation at Glacier Bay seems to have been due mainly to different rates of dispersal, growth and reproduction, and to different life spans, among the invading species. Species with especially mobile propagules, e.g. spores of *Rhacomitrium* and *Equisetum*, plumed seeds or fruits of *Epilobium*, *Dryas* and *Salix*, arrived in the greatest number and so formed the bulk of the early vegetation. Shrubs and conifers invaded more slowly, though pro-

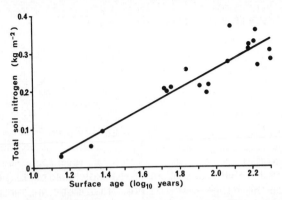

Fig. 5.2 Accumulation of nitrogen in the soil with time at the recessional moraines of the Mendenhall and Herbert glaciers, Alaska. (After Crocker and Dickson 1957 [247].)

gressively, eventually overtopping the earlier vegetation, while the shrubs were themselves later overtopped by conifers, dying of shading or old age and unable to regenerate under the conifer canopy. The process of succession seems identical to that of secondary succession as explained by Egler [50] and Drury and Nisbet [1] (see Section 1.3.2). Any glacier seems to control the degree to which succession on adjacent bare ground can advance through its effect on the local microclimate, but this control seems effective only over very short distances [244, 245].

Studies of soil development behind retreating glaciers have shown that quite considerable changes occur during vegetation succession [246, 247, 248]. Typically, organic matter accumulates on and in the soil, with a corresponding increase in soil nitrogen content, while pH declines as bases are leached out of the mineral soil and as organic debris produces cation exchange sites which are occupied by hydrogen ions in the absence of bases. These changes tend to show a logarithmic relationship with time; Fig. 5.2 gives an example.

Nitrogen accumulates particularly under *Dryas* and *Alnus*: both have root nodules which appear to fix atmospheric nitrogen [246, 247]. It has been claimed that pioneering shrubs and trees at Glacier Bay are slow growing because of nitrogen deficiency, but that when growing in association with *Dryas* and *Alnus* their growth is stimulated [241, 249]. This is an interesting autogenic effect, but there does not seem to be conclusive evidence that *Dryas* and *Alnus* are essential to the succession. Trees invading non-forested areas can suffer a lag before vigorous growth begins because of a delay in infection by mycorrhizal fungi, spores of which may also have to colonize. For example, *Pinus sylvestris* seedlings often show such a check in growth when invading *Calluna*-dominated moorland in Britain.

6 Secondary successions

The subject of secondary succession has already been variously touched upon in the preceding chapters. It was defined as the changes in composition that ensue when a patch of vegetation suffers gross disturbance such that many or all of the individual plants present are killed. When this happens, species other than those previously present commonly establish and occupy the gaps. If, as frequently happens, the invaders comprise species showing wide variations in longevity, pronounced changes will occur in the composition of the vegetation as short-lived species die, and the new gaps are re-occupied, while longer-lived species remain.

Such changes occur because modern floras contain species with a great variety of ecological strategies, adapted to exploit different spatiotemporal niches. To understand why, it is first necessary to grasp just how unstable the natural environment is, and probably always has been, during the course of evolution. Many environmental factors regularly disrupt vegetation in a catastrophic manner. Some were noted at the beginning of Chapter 2, and the list could be expanded. Several factors may operate in a given area. Thus, from analysis of old land surveyors' records in part of Wisconsin, Stearns [67] concluded 'Windfall alone or with the other agents of mortality, fire, drought, glaze storm, insect or fungus infestation and senescence keeps the forest in a constant state of flux'. Similar conclusions have been reached in studies on forest elsewhere in North America [250, 251, 252].

Wind, and fire caused by lightning, are particularly widespread and important agents of catastrophic disturbance. Wind-throw is often important in temperate and boreal forests [99, 100, 196, 253], and in many tropical rain forests [23, 101, 102]. Fire is of widespread occurrence in forests [57, 96, 134, 254, 255], and even some tropical rain forests may burn [23, 97]. Many forests seem to have been burnt naturally as often as every 50 to 100 years [57, 256, 257].

Tropical rain forest has often been thought of as a very stable, unchanging vegetation with a very predictable environment. However, it does not seem to endure unchangingly through time any more than any other kind of vegetation [70, 258, 259]. The apparent environmental predictability now seems less certain. Even in forests with no dry season, particularly heavy rains can occur at random intervals throughout the year and cause normally quiet streams to flood the forest floor [260].

The other major world classes of vegetation are grassland, and the vegetation of deserts and tundra. Grassland seems to have suffered periodic catastrophic disturbance as much as forest, in particularly by fire and drought [54, 134, 135], while desert vegetation suffers from recurring drought [267], and that of tundra from constant physiographic instability [94, 95].

The evidence thus suggests that most vegetation has historically undergone periodic gross disturbance. This has probably also been a feature of the environment throughout much of evolutionary time: forest fire certainly occurred at least as far back as the Mesozoic, and may have been common in the Tertiary when today's flora largely evolved [261].

Plants today exhibit a wide variety of ecological strategies which have presumably evolved to take maximum advantage of the niche diversity created when environmental instability keeps vegetation in a state of flux. Two contrasting trends in strategy can be recognized in particular. One is of the fast-growing, shade intolerant herb, which devotes a large proportion of its biomass to producing large numbers of small, long-lived and/or widely dispersed seeds, often in its first growing season. Such a plant is well adapted to exploit temporary gaps in vegetation. It will colonize rapidly, either from having persisted between disturbances as viable seed buried in the soil, or because it is a ubiquitous component of the seed rain. When established it will quickly complete its life cycle and produce a new crops of seeds. The cosmopolitan weed *Capsella bursa-pastoris*, and the herbaceous *Senecio* spp., are good examples. The contrasting extreme is that of the slow-growing, long-lived tree, which devotes most of its biomass to producing large stature, has seedlings tolerant of shade, and produces relatively fewer, larger, shorter-lived and often poorly dispersed seeds. Such a plant is adapted to a stable environment, being able to monopolize light and space for long periods, and to establish young plants in shade. Northern Hemisphere examples include sugar maple (*Acer saccharum*), the hemlocks, *Tsuga* spp., and the beeches, *Fagus* spp.

Organisms showing these two contrasting tendencies have been termed r- and K-strategists respectively, after equations in population dynamics, where r represents the intrinsic rate of increase of a population when growth is not limited by resources, and K the population density at which a population is in equilibrium with the resources [121]. In practice, however, any species possesses one of a very large number of combinations of attributes, and most have combinations intermediate between the two extremes. The terms thus have meaning only when species are compared. A short-lived tree producing large numbes of small, efficiently dispersed seeds, such as most species of *Betula*, is more of an r-strategist than a species of *Fagus*, but more of a K-strategist than most herbs. Species with a clear-cut r-strategy have been termed plant nomads or vagrants, pioneers, or 'early successional' species. Species with a pronounced K-strategy are those that have traditionally been thought of as 'climax' species. The r and K terminology, although currently in vogue, is of rather limited usefulness in the context of vegetation dynamics. It is possible and useful to recognize further arbitrarily chosen strategies between these two extremes, or different kinds of strategy. Budowski [262] divided tropical American rain forest species into four classes: pioneer, early secondary, late secondary and climax. Whitmore [23, 102] divided the top-of-canopy tree species in Solomon Islands rain forest into four groups on the basis of their response to and need for canopy gaps in regeneration. Grime [263] has classified herbaceous species by the relative importance of their strategies towards stress, disturbance, and competition.

In summary, it seems that environmental instability and frequent disturbance of vegetation in the past have resulted in the evolution of modern floras that embrace a great variety of niches or ecological strategies. This variety now ensures the changes seen during secondary succession after vegetation has been disturbed.

6.1 The course of secondary succession

It was noted in Section 3.1 that on a scale larger than that of the patch undergoing secondary succession, a forest may appear as mosaic or cyclic series of different phases of regeneration. Fig. 6.1 shows an example, an area of rain forest with different phases of canopy regrowth in storm and clearance gaps. In the sense that succession on the individual patch often eventually reproduces a similar vegetation to that originally disturbed, this process also may appear cyclic. However, in view of the long duration of many successions, and the often doubtful extent to which vegetation is ever reproduced in time, the term is not generally helpful.

Secondary succession typically appears as a series of changes in species populations with different species or groups of species successively attaining and then losing predominance. Clements [2] attributed these changes to successive influxes of colonizing species, a view long accepted, despite accumulating evidence to the contrary. Egler in 1953 [50] seems to have been the first seriously to challenge its generality (see Section 1.3.2). It now seems clear that most species seen during the majority of secondary successions are either present at the outset as buried viable seed or other

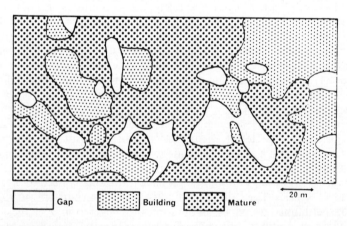

Fig. 6.1 Canopy phases in 2 ha of rain forest at Sungei Menyala, Malaya, 1971, resulting from windfall, and, on the right, from partial clearance in 1917. (From *Tropical Rain Forests of the Far East* by T.C. Whitmore, published by Oxford University Press [23].)

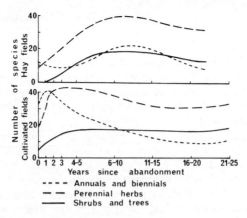

Fig. 6.2 Variation in numbers of species in abandoned fields of different ages in Michigan. (After Beckwith, 1954 [264].)

plant parts, or colonize within the first few years. Fig. 6.2 illustrates this for abandoned fields in Michigan. It shows (a) that a large number of species were present at the outset, (b) that colonization was rapid during the first ten years, and (c) that shrubs and trees established early on, concurrently with herbaceous species.

Changes in composition and appearance of vegetation with time during succession are largely caused by differences in growth and survival rates, competitive ability and longevity. This is illustrated by Fig. 6.3, which shows some typical changes following disturbance of northern hardwood forest in New Hampshire, with fast-growing, shade intolerant species gradually being replaced in importance by the long-lived, slow-growing, shade-tolerant *Acer saccharum* and *Fagus grandifolia*. Successions at actual sites are more complex and variable, but show the same trends towards increasing importance of shade tolerant species [266, 267].

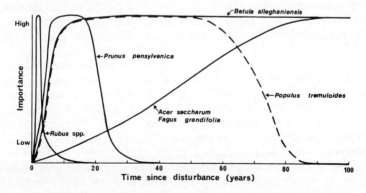

Fig. 6.3 Diagrammatic representation of the changing importance of different species with time following disturbance of a typical northern hardwood forest in New Hampshire. (From Marks, 1974 [265], copyright by the Ecological Society of America.)

6.2 Factors determining the course of secondary succession

6.2.1 The local vegetation
The potential maximum duration and complexity of secondary succession is determined by the species richness of the local flora. The simplest successions occur in desert and tundra vegetation, where succession is usually a matter of simple re-invasion by the original species as no others may exist [268, 269, 270, 276].

The importance of the residual vegetation and buried seed population in influencing the course of succession at disturbed sites has often been demonstrated [263, 271–275]. However, the buried seed population never seems to include seeds of long-lived 'climax' species [52, 275], which have a characteristically short period of viability [262, 273, 274], and therefore need to seed in unless young plants are present at the time of disturbance. Frequent disturbance reduces the species richness of the residual vegetation and the buried seed flora [273, 277]. The same effect would be expected after long periods without disturbance, when the seed store of 'successional' species would not be replenished, but this does not seem to have been examined.

6.2.2 Characteristics of the disturbed site
The size of an opening or gap made in a closed canopy, and the nature and degree of disturbance, are important. Some species need a certain minimum gap size for establishment [129, 278]. In forest, the size of gap determines whether 'successional' or 'climax' species colonize it [102, 279–282]: Kramer [279] found that in rain forest in Java, the shade tolerant 'climax' dominants regenerated in gaps less than about 0.1 ha, but that in gaps greater than 0.2–0.3 ha, they were completely suppressed by light demanding 'successional' species. The degree of ground disturbance controls the extent to which buried seed is exposed to conditions suitable for germination and subsequent establishment, and also to which suitable conditions are created for the establishment of different incoming species (see Section 2.3). Fig. 6.2 shows that the rate of establishment of new species during an old-field succession was much faster in arable fields than in meadow, where there was already complete cover.

The availability of seed sources is another important determinant of succession. The rate of succession on abandoned fields has often been related to the distance from stands of natural or semi-natural vegetation [277, 283, 284]. In the Scottish Highlands, most *Calluna* moorland, which was formerly natural woodland, tends simply to regenerate itself after disturbance just because sources of tree seed are so scarce [285].

Species colonizing a disturbed patch tend to show differences in distribution corresponding to differences in soil conditions, which may be intrinsic to the site [286] or caused by the disturbance. Thus it was found that when trees were uprooted by wind in one New Hampshire forest, *Betula lenta* tended to establish on the windfall mounds, and *Acer rubrum* in the pits [58].

6.2.3 Stable vegetation

During secondary succession, herbaceous or shrubby vegetation may develop which changes little over many years, effectively resisting invasion by trees. Some patches are known to have been stable for 50 years or more [287]. The grass balds of the Southern Appalachian Mountains seem to have been present at the time of white settlement, and their mention in Cherokee legend suggests an even greater age [288]. Most reports are from North America [31, 287–290], but this probably just reflects a greater interest there in the phenomenon. The grass *Imperata cylindrica* often forms dominant stands in the Old World tropics [22], and can resist invasion by trees [291]. The cosmopolitan fern *Pteridium aquilinum* is also often found in dominant stands [22, 31, 142, 292], which in some cases seem to resist tree invasion indefinitely.

Little is known of why such patches remain stable. Certainly, shade cast by the often dense canopies will tend to prevent early seedling establishment, and with species like *Pteridium* the smothering effect of litter is probably also important. Allelopathic effects may contribute to the stability of patches of *Andropogon virginicus* [289] and *Pteridium* [292, 293], and similar evidence may be found for other species if it is sought.

6.2.4 Site modification

The previous section showed ways in which site modification may retard succession. Facilitation of succession was discussed in Section 2.5; a further example is of those 'climax' species that regenerate only in small gaps because of competition from faster-growing but shade-intolerant species in larger gaps. Allelopathy may also facilitate succession. The rapid disappearance of pioneer weeds in early old-field successions in parts of the United States seems in many instances to be promoted by the production of toxins that inhibit growth [139].

Many soil properties, in particular those associated with nutrient cycling, inevitably change during succession, but there is little evidence that they materially influence the species changes. They often do however, when *Calluna* moorlands in Britain are invaded by *Betula pendula* or *B. pubescens*. Gross changes can then occur in soil pH and nutrient availability, and many field layer species characteristic of woodland, which could not grow in the moorland soil, often appear. At one site the number of vascular species in the field layer changed from 12 on the *Calluna* moorland to 24 under a 38-year-old stand of *Betula pendula* [231].

6.2.5 The role of chance

Succession shows an endless variety of patterns—no two successions have yet been recorded with identical mixtures and proportions of species. Many workers have stressed the role of chance in causing this variety [271, 272, 274, 280, 281, 294]. Chance factors include variations in the occurrence, severity and nature of disturbance, in the occurrence of good seed years for particular species and of favourable weather conditions for establishment, and in the vagaries of seed dispersal. Few factors influencing succession operate in a deterministic way (a notable exception being the patterns imposed on vegetation by dominant species), but neither do they seem to operate truly at random. Instead, the chance working of most can be treated in a probabilistic manner [274].

6.3 Predictability of secondary succession
In recent years there have been increasing attempts to predict the outcome of succession using mathematical models. These have been summarized by Slatyer [59]. The probabilistic models using Markov chains are particularly interesting, especially the plant-by-plant replacement models of Horn [266, 295, 296]. However, current models reflect few of the complexities of succession. While model-building seems likely to continue as a major growth point in research on vegetation dynamics, more sophisticated mathematical techniques and more detailed accounts of actual successions are needed if the past rate of progress is to continue.

7 Changes caused by grazing animals

In the last four chapters almost no mention was made of the effects of herbivores. However, grazing and browsing by domestic stock and wild herbivores has a controlling influence on a large part of the vegetation of the world's uncultivated land. Invertebrate herbivores tend to be more specialized feeders than vertebrate herbivores. Though they can have profound effects on the population dynamics of individual species [297, 298], they do not seem able to control the physiognomy and life form of vegetation in the way that many vertebrate herbivores can [109, 110].

Grasslands are the prime example of vegetation often controlled, and even created, by grazing. The basal meristems of typical grassland species presumably evolved as an adaptation to grazing. One result of heavy grazing pressure is to reduce the vigour of potential dominants and thus maintain the species diversity of swards. Fig. 7.1 gives an example. It shows the changes that followed exclusion of rabbits from an area of

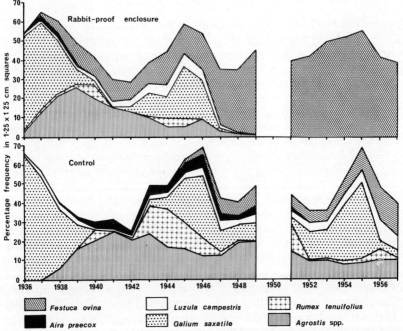

Fig. 7.1 Changes in species composition of a grassland inside and outside a rabbit-proof enclosure in the English Breckland. (From Watt, 1960 [220], courtesy of *Journal of Ecology*.)

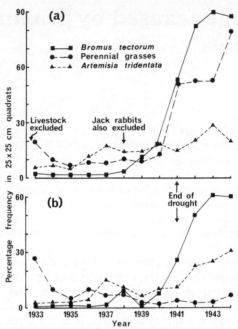

Fig. 7.2 Changes in abundance of major components of (a) fenced, and (b) unfenced rangeland in southern Idaho. (After Piemeisel, 1945 [299].)

grassland in the English Breckland. In the enclosure the grass *Festuca ovina* gradually increased in abundance, became dominant after 11 years, and eventually caused the extinction of all other species except for a few shoots of *Agrostis tenuis*. The unfenced control plot is a further example of just how marked year-to-year fluctuations in composition can be (see Chapter 3). Also, even after *F. ovina* became dominant in the enclosure, fluctuations in the total amount of vegetation present closely followed those on the control plot, indicating control by extrinsic factors. (Spring rainfall appeared to be important.)

Fig. 7.2 indicates the effects of grazing and drought on rangeland in southern Idaho. In 1933 a plot was fenced against cattle, and then against jack rabbits (*Lepus* spp.) also after the 1938 growing season. However, vegetation in the fenced plot did not differ appreciably from the control until 1941, the year in which the great 1930's drought ended (see Fig. 3.2). There was then a dramatic increase in the fenced plot in the abundance of perennial grasses (which are preferred forage), but not in the control, while the less attractive *Bromus tectorum* and *Artemisia tridentata* increased in both fenced and control plots. This suggests that while drought limited the abundance of all species, grazing was only controlling the perennial grasses. Other evidence suggested that jack rabbits rather than cattle were largely responsible for the grazing effect.

Controlling the grazing of domestic stock is the most effective way of controlling the floristic composition of grazing lands. Table 7.1 shows

Table 7.1 Changes in composition (by % weight of herbage) of an upland grassland on mineral soil in Wales after protection from grazing by domestic stock (From Jones, 1967 [300], courtesy of the Welsh Plant Breeding Station.)

	1930 Prior open grazing	1932 After 2 years without grazing	1944 After 14 years without grazing	1956 After 12 years subsequent open grazing
Agrostis spp.	9	2	Trace	25
Calluna vulgaris	Trace	20	85	5
Festuca ovina	80	60	5	38
Vaccinium myrtillus	1	10	8	10
Other species	10	8	2	22

how profound changes can occur; it is an extreme example as grazing by stock was completely prevented. In this sward the dwarf shrub *Calluna vulgaris*, which initially comprised less than 1% of the forage, had increased to 20% after only two years without grazing. After 14 years it formed 85% of the vegetation, nearing complete dominance. However, these changes were reversible. After 12 years of subsequent open grazing, *Calluna* had declined to only 5%.

Generalizations about likely changes with grazing can be made for particular regions. Fig. 7.3 shows the generalized successional relationships between the four most abundant vegetation types of well-drained, acid soils in upland Britain. With very low grazing pressures all types tend to revert to woodland, grassland often via a *Calluna* phase when the rate of tree invasion is slow. *Pteridium* may invade grassland and heath and patches often resist tree invasion. Woodland often seems to regenerate via grassland or heath phases, especially with increasing

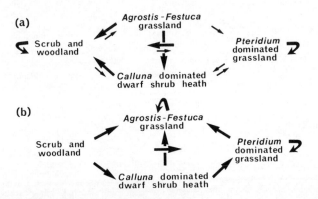

Fig. 7.3 Generalized sequences of successional relationships between four widespread types of vegetation of well-drained, acid soils in upland Britain, given (a) low or negligible, and (b) high grazing pressures. (Large arrows represent usual sequences, small arrows less common sequences.) (After Miles, 1979 [301].)

altitude and latitude [302]. With fairly high grazing pressures, i.e. greater then about 2.5 sheep per hectare or equivalent, the overall trend is towards grassland formation. Regeneration by trees is prevented, while dwarf shrubs are either killed by mechanical damage or are grazed down to a level at which they are at a competitive disadvantage against herbaceous species. *Pteridium* patches may still remain however. They are only lightly grazed, but whereas cattle will trample and kill the fronds, sheep tend to walk between them.

8 Concluding remarks

We have seen that change in time is ubiquitous. Vegetation change stems from many sources: the seasonal development of component species, processes of self-regeneration, the life-cycles of dominant species, variations in climate (year-to-year or over longer periods), evolutionary developments, and gross disturbance to existing vegetation or the exposure of ground by natural physiographic processes (when succession results). These changes occur concurrently, superimposed on endless variation in composition in space. The result is that patterns of vegetation change in time show a bewildering variety.

The study of vegetation presents a dilemma which is common to many branches of biological science. On the one hand, vegetation shows endless variation in composition in time and space. Hence any classification of it has to use arbitrary criteria, and the different units thus identified inevitably intergrade. On the other hand, in order to study vegetation, or any other biological phenomena, it is necessary to create order, to identify small units which it is possible to study. It is important to recognize that any classification is only a working hypothesis, an ad hoc fiction necessary to advance scientific understanding, but whose usefulness is limited to the particular situation for which it was formulated. Unfortunately, the essential purpose of classification and its intrinsic limitations seem often to have been overlooked.

In Chapter 1, I unequivocally rejected the organismic view of vegetation, though I accept that, as a result of species interactions, vegetation may arguably be regarded as possessing its own emergent properties. Thereafter, in dealing with vegetation change, I had to write as though discrete and real vegetation types existed. This might suggest that I in fact use the concept I rejected. I do not. Any terms which may imply this are just simplifications, and hence fictions, needed for communication.

I discussed models of vegetation change, and concluded that no one model operates exclusively, and that different models may operate sequentially or concurrently at the same place. These models have been discussed in more detail by Connell and Slatyer [303]. I cannot put forward an alternative, all-embracing hypothesis of my own—in fact I doubt if this will ever be possible. Explanations of vegetation change are found at the level of the species and the individual. Vegetation composition is the result of the interactions of species with varying ecological tolerances and requirements. Species attributes of particular importance are dispersal mechanisms, size and duration of viability of seeds, growth rates, life form and plant size, longevity, and tolerance to environmental stresses at different phases of the life cycle. Changes in the physical or biotic environment alter any balance of interactions that may have existed, and cause changes in vegetation with time. Patterns of change in time and space are inseparably linked; each influences the other.

The terminology of vegetation dynamics is confused. This was inevitable in the historical context. Vegetation is a concept covering the varied assemblages of plants found from place to place. Vegetation change is a process involving changes in the populations of species. Any attempts to define the latter at the different conceptual level of the former can only produce unsatisfactory, intergrading categories. However, as yet there is no alternative terminology, so vegetation-level terms must be used. The distinctions between regeneration, fluctuation and succession, though arbitrary, are generally useful. It would seem desirable to discard most other terms, e.g. the distinction between primary and secondary succession. Instead, a brief, explicit description should be given of each instance of vegetation change studied e.g. changes (or succession) following forest fire, changes (or fluctuations) with drought.

Our understanding of vegetation dynamics is still rudimentary. More information is needed at every level. For example, competition is still poorly understood in the field. Allelopathy may be of considerable significance, but evidence for it is largely circumstantial and often weak. There is a particular dearth of long-term studies of vegetation change. These are needed for the insights they may provide about patterns and processes of change, and for information on the incidence, nature and role of natural disturbances. It seems that long term stability in vegetation may be the exception, and episodic disturbance the rule. Short term studies are unlikely to detect the rare event which may determine vegetation pattern for decades or centuries. At worst, they may only produce 'cursory correlations of the coincidentally concomitant' [29]!

References

References marked with an asterisk (*) in this list are particularly recommended for further study. They are briefly annotated to aid selection of papers on particular topics, and selection within topics where more than one reference is recommended.

[1] *Drury, W.H. and Nisbet, I.C.T. (1973), *J. Arnold Arbor.*, **54**, 331–368.
The most detailed critique available of Clementsian and neo-Clementsian ideas. Essential reading for anyone interested in succession.

[2] *Clements, F.E. (1916), *Carnegie Inst. Wash. Publ.* No. 242.
Valuable descriptions of early studies of vegetation dynamics and of the causes of change, but his model of succession does not fit the facts. A condensed version (Clements, F.E. (1928), Plant Succession and Indicators, Wilson, New York) has recently been reprinted by Hafner, New York.

[3] Carpenter, J.R. (1938), *An Ecological Glossary*, University of Oklahoma Press, reprinted by Hafner, New York.

[4] Hanson, H. (1962), *Dictionary of Ecology*, Peter Owen, London.

[5] *Raup, H.M. (1975), *J. Arnold Arbor.*, **56**, 126–163.
An 'eye-opener' for all addicts of vegetation classification.

[6] Gleason, H.A. (1926), *Bull. Torrey Bot. Club*, **53**, 7–26.

[7] Braun-Blanquet, J. (1932), *Plant Sociology*, McGraw-Hill, New York.

[8] *Gleason, H.A. (1939), *Am. Midl. Nat.*, **21**, 92–110.
A repetition and elaboration of the 1926 paper. Highly recommended.

[9] Watt, A.S. (1964), *J. Ecol.*, **52**(Suppl.), 203–211.

[10] Lenoble, F. (1926), *Bull. Soc. Bot. France*, **73**, 873–893.

[11] Lenoble, F. (1928), *Archs Bot. Bull. mens.*, **2**, 1–14, 129–131.

[12] Ramensky, L.G. (1926), *Bot. Cbl.*, **7**, 453–455.

[13] Ramensky, L.G. (1930), *Beitr. Biol. Pfl.*, **18**, 269–304.

[14] Goff, F.G. and Zedler, P.H. (1968), *Ecol. Monogr.*, **38**, 65–86.

[15] Mason, H.L. (1947), *Ecol. Monogr.*, **17**, 201–210.

[16] West, R.G. (1964), *J. Ecol.*, **52**(Suppl.), 47–57.

[17] Watts, W.A. (1973), in *Quarternary Plant Ecology*, (Ed.) H.J.B. Birks and R.G. West, pp. 195–206, Blackwell, Oxford.

[18] Godwin, H. (1975), *History of the British Flora*, 2nd edn, Cambridge University Press, Cambridge.

[19] Poore, M.E.D. (1955), *J. Ecol.*, **43**, 245–269.

[20] Whittaker, R.H. (Ed.) (1973), *Ordination and Classification of Communities*, W. Junk, The Hague.

[21] Levins, R. (1975), in *Evolution and Ecology of Communities*, (Ed.) M.L. Cody and J.M. Diamond, pp. 16–50, Harvard University Press, Cambridge, Mass.

[22] Richards, P.W. (1952), *The Tropical Rain Forest*, Cambridge University Press, Cambridge.

[23] Whitmore, T.C. (1975), *Tropical Rain Forests of the Far East*, Oxford University Press, Oxford.

[24] Kozlowski, T.T. (1971), *Growth and Development of Trees*, Vol. II, Academic Press, New York.

[25] Bormann, F.H. (1966), *Ecol. Monogr.*, **36**, 1–26.

[26] *Whittaker, R.H. and Woodwell, G.M. (1972), in *Ecosystem Structure and Function*, (Ed.) J.A. Wiens, pp. 137–159, Oregon State University Press, Corvallis.
A good discussion of vegetation properties.

[27] Whittaker, R.H. (1975), in *Unifying Concepts in Ecology* (Ed.) W.H. van Dobben and R.H. Lowe-McConnell, pp. 169–181, W. Junk, The Hague.

[28] Harper, J.L. (1964), *J. Ecol.*, **52**(Suppl.), 149–158.

[29] *Egler, F.E. (1977), *The Nature of Vegetation. Its Management and Mismanagement*, Aton Forest, Norfolk, Conn.
An unorthodox presentation, but an extremely valuable book. Includes sections on the nature of vegetation and vegetation dynamics.

[30] Yapp, R.H., John, D. and Jones, O.T. (1917), *J. Ecol.*, **5**, 65–103.

[31] Curtis, J.T. (1959), *The Vegetation of Wisconsin*, University of Wisconsin Press, Madison.

[32] Whittaker, R.H. (1967), *Biol. Rev.* **49**, 207–264.

[33] Whittaker, R.H. (1975), *Communities and Ecosystems*, 2nd edn, Macmillan, New York.

[34] *Knapp, R. (Ed.) (1974), *Vegetation Dynamics*, W. Junk, The Hague.
A good source of references, with a few good papers, but most papers are disappointing.

[35] Rabotnov, T.A. (1974), in *Vegetation Dynamics* (Ed.) R. Knapp, pp. 19–24, W. Junk, The Hague.

[36] Davies, M.S. and Snaydon, R.W. (1973), *J. appl. Ecol.*, **10**, 33–45.

[37] Spurr, S.H. (1952), *Ecology*, **33**, 426–427.

[38] Cowles, H.C. (1899), *Bot. Gaz.*, **27**, 95–117, 167–202, 281–308, 361–391.

[39] Cowles, H.C. (1901), *Bot. Gaz.*, **31**, 73–108, 145–182.

[40] Cowles, H.C. (1911), *Bot. Gaz.*, **51**, 161–183.

[41] Clements, F.E. (1904), *Rep. Bot. Surv. Nebraska*, No. 7.

[42] Clements, F.E. (1936), *J. Ecol.*, **24**, 252–284.

[43] Gleason, H.A. (1917), *Bull. Torrey Bot. Club*, **53**, 7–26.

[44] *Gleason, H.A. (1927), *Ecology*, **8**, 299–326.
Develops the argument of [8] in relation to succession. Recommended reading.

[45] Collier, B.D., Cox, G.W., Johnson, A.W. and Miller, P.C. (1973), *Dynamic Ecology*, Prentice-Hall, Englewood Cliffs, N.J.

[46] Emlen, J.M. (1973), *Ecology, an Evolutionary Approach*, Addison-Wesley, Reading, Mass.

[47] Kershaw, K. (1974), *Quantitative and Dynamic Plant Ecology*, 2nd edn, Edward Arnold, London.

[48] Ricklefs, R.E. (1973), *Ecology*, Nelson, London.

[49] *Colinvaux, P.A. (1973), *Introduction to Ecology*, Wiley, New York.
A good introduction to plant succession. Recommended reading.

[50] *Egler, F.E. (1954), *Vegetatio*, **4**, 412–417.
A milestone in our understanding of succession.

[51] Ødum, S. (1965), *Dansk Bot. Arkiv*, **24**, No. 2.

[52] *Livingston, R.B. and Allessio, M.L. (1968), *Bull. Torrey Bot. Club*, **95**, 56–69.
An example of how forest may contain all its successional precursors as viable buried seed.

[53] Harper, R.M. (1914), *Bull. Torrey Bot. Club*, **41**, 209–220.
[54] *Coupland, R.T. (1958), *Bot. Rev.*, **24**, 273–317.
      A good summary of the dynamics of the Great Plains grasslands.
[55] Küchler, A.W. (1972), *Erdkunde*, **26**, 120–129.
[56] Louks, O.L. (1970), *Am. Zool.*, **10**, 17–25.
[57] *Wright, H.E. and Heinselman, M.L. (Ed.) (1973), *Quaternary Res.*, **3**, 317–513.
      An excellent series of papers from a symposium on the role of fire in North American conifer forests.
[58] *Henry, J.D. and Swann, J.M.A. (1974), *Ecology*, **55**, 772–783.
      An intriguing investigation of the past dynamics of a New Hampshire forest. Highlights the importance of natural disturbances.
[59] *Slatyer, R.O. (Ed.) (1977), *Dynamic Changes in Terrestrial Ecosystems: Patterns of Change, Techniques for Study and Applications to Management.* UNESCO, Paris.
      Discusses recent attempts to make mathematical models of successions.
[60] *McCormick, J. (1968), *Via* (Student Publication, Graduate School of Fine Arts, University of Pennsylvania), **1**, 22–35, 131–132.
      A comprehensive and readily comprehensible account and critique of Clements' model of succession.
[61] *Drury, W.H. and Nisbet, I.C.T. (1971), *General Systems*, **16**, 57–68.
      Discusses what may have been a basic misunderstanding underpinning Clements' model of succession.
[62] Ratcliffe, D.A. (Ed.) (1977), *A Nature Conservation Review*, Cambridge University Press, Cambridge.
[63] Tansley, A.G. (1935), *J. Ecol.*, **16**, 284–307.
[64] Margalef, R. (1963), *Am. Nat.*, **97**, 357–374.
[65] Margalef, R. (1968). *Perspectives in Ecological Theory*, University of Chicago Press, Chicago.
[66] *Odum, E.P. (1969), *Science*, **164**, 262–270.
      An interesting attempt to produce a model of succession based on changes expected in ecosystem 'properties'.
[67] Stearns, F.W. (1949), *Ecology*, **30**, 350–358.
[68] Peterken, G.F. (1976), *J. Ecol.*, **64**, 123–146.
[69] *Hastings, J.R. and Turner, R.M. (1965), *The Changing Mile*, University of Arizona Press, Tucson.
      An outstanding photographic documentation of vegetation change in southern Arizona and northwest Mexico.
[70] Kendall, P.L. (1969), *Ecol. Monogr.*, **39**, 121–176.
[71] Iversen, J. (1969), *Oikos*, Suppl. **12**, 35–49.
[72] Dimbleby, G.W. (1962), *Oxf. For. Mem.* No. 23.
[73] Havinga, A.J. (1973), *Meded. Landb Hogesch. Wageningen*, **63**, no. 1.
[74] *Swain, A.M. (1973), *Quaternary Res.*, **3**, 383–396.
      Fascinating use of charcoal in lake sediments to study the importance of fire as a natural factor in the forest dynamics of northeastern Minnesota.
[75] Romans, J.C.C. and Robertson, L. (1975), in *The Effect of Man on the Landscape: the Highland Zone*, (Ed.) J.G. Evans, S. Limbrey and H. Cleere, pp. 37–39, Council for British Archaeology, London.
[76] Witty, J.E. and Knox, E.G. (1964), *Proc. Soil Sci. Soc. Amer.*, **28**, 685–688.
[77] Guevara, S. and Gomez-Pompa, A. (1972), *J. Arnold Arbor.*, **53**, 312–335.
[78] *Olson, J.S. (1958), *Bot. Gaz.*, **119**, 125–170.
      An excellent study of changes in vegetation and soils on the southern Lake Michigan sand dunes.

[79] Cooper, W.S. (1923), *Ecology*, **4**, 223–246, 355–365.

[80] Cooper, W.S. (1931), *Ecology*, **12**, 61–65.

[81] Cooper, W.S. (1939), *Ecology*, **20**, 130–159.

[82] Tansley, A.G. (1939), *The British Isles and their Vegetation*, Cambridge University Press, Cambridge.

[83] *Walker, D. (1970), in *Studies in the Vegetational History of the British Isles*, (Ed.) D. Walker and R.G. West, pp. 117–139, Cambridge University Press, Cambridge.
      A new classic which should be compulsory reading for everyone interested in succession.

[84] Ratcliffe, D. (1961), *J. Ecol.*, **49**, 187–203.

[85] LaMarche, V.C. (1969), *Ecology*, **50**, 53–59.

[86] Johnson, F. and Raup, H.M. (1947), *Pap. Robert S. Peabody Fdn Archaeol.*, **1**, No. 1.

[87] Heath, J.P. (1967), *Ecology*, **48**, 270–275.

[88] Johnson, W.C., Burgess, R.L. and Keammerer, W.R. (1976), *Ecol. Monogr.*, **46**, 59–84.

[89] Viereck, L.A. (1970), *Arctic Alpine Res.*, **2**, 1–26.

[90] Ranwell, D.S. (1964), *J. Ecol.*, **52**, 79–94.

[91] Viereck, L.A. (1966), *Ecol. Monogr.*, **36**, 181–199.

[92] Tagawa, H. (1964), *Mem. Fac. Sci. Kyushu Univ., Ser. E Biol.*, **3**, 166–228.

[93] Scifres, C.J. and Mutz, J.L. (1975), *J. Range Mgmt*, **28**, 279–282.

[94] *Raup, H.M. (1951), *Ohio J. Sci.*, **51**, 105–116.
      A good account of the role of physiographic processes in boreal regions.

[95] *Drury, W.H. (1956), *Contr. Gray Herb. Harv. Univ.* No. 178.
      An excellent study relating vegetation changes and physiographic processes in Alaska, with a particularly useful discussion.

[96] Maissurow, D.K. (1941), *J. For.* **39**, 201–207.

[97] Ridley, W.F. and Gardner, A. (1961), *Aust. J. Sci.*, **23**, 227–228.

[98] *Komarek, E.V. (1969), *Proc. Ann. Tall Timbers Conf.*, **8**, 169–197.
      An introduction to the ecological effects of fire and lightning.

[99] *Sernander, R. (1936), *Acta phytogeogr. suec.* No. 8.
      Discusses the role of windfall in the regeneration of Swedish spruce forest.

[100] *Lutz, H.J. (1940), *Yale Univ. Sch. For. Bull.* No. 45.
       Discusses the effects of windfall in New Hampshire forests.

[101] *Webb, L.J. (1958), *Aust. J. Bot.*, **6**, 220–228.
       Discusses the role of cyclones in tropical rain forest in Queensland.

[102] *Whitmore, T.C. (1974), *Commonw. For. Inst. Pap.* No. 46
       Discusses the role of cyclones on tropical forest in the Solomon Islands.

[103] Watson, A., Miller, G.R. and Greene, F.H.W. (1966), *Trans. Proc. bot. Soc. Edinb.*, **40**, 195–203.

[104] *Sprugel, D.G. (1976), *J. Ecol.*, **64**, 889–911.
       An interesting account of 'wave regeneration' in a balsam fir forest.

[105] Billings, W.D. (1969), *Vegetatio*, **19**, 192–207.

[106] Raup, H.M. (1937), *J. Arnold Arbor.*, **18**, 79–117.

[107] *Kulman, H.M. (1971), *A. Rev. Ent.*, **16**, 289–324.
       A good reference source on the effects of insects on trees.

[108] *Gray, B. (1972), *A. Rev. Ent.*, **17**, 313–354.
       A reference source on the effects of insects on tropical tree species.

[109] Nicholson, I.A. (1970), *Trans. Proc. bot. Soc. Edinb.*, **41**, 85–94.

[110] Welch, D. (1974), *Land*, **1**, 59–68.

[111] Klinkowski, M. (1970), *A. Rev. Phytopathol.*, **8**, 37–60.

[112] Sarvas, R. (1948), *Comm. inst. for. Fenn.*, **35**, no. 4.

[113] Maguire, B. (1963), *Ecol. Monogr.*, **33**, 161–185.

[114] Harberd, D.J. (1961), *New Phytol.*, **60**, 184–206.

[115] Fox, J.E.D. (1972), *The Natural Vegetation of Sabah and Natural Regeneration of the Dipterocarp Forests.* Unpublished Ph.D. Thesis, University of Wales, Bangor.

[116] Brereton, A.J. (1971), *J. Ecol.*, **59**, 321–338.

[117] Godwin, H. (1923), *J. Ecol.*, **11**, 160–164.

[118] Morrison, R.G. and Yarranton, G.A. (1973), *Can. J. Bot.*, **6**, 172–182.

[119] Docters van Leeuwen, W.M. (1936), *Ann. Jard. Bot. Buitenzorg.*, **46–47**, 1–507.

[120] Einarsson, E. (1967), *Aquilio, Ser. Bot.*, **6**, 172–182.

[121] MacArthur, R.H. and Wilson, E.O. (1967), *The Theory of Island Biogeography*, Princeton University Press, Princeton.

[122] *Grubb, P.J. (1977), *Biol. Rev.*, **52**, 107–145.
      A scholarly review of the processes of plant regeneration in which the idea of the regeneration niche is advanced and its importance stressed.

[123] Janzen, D.H. (1971), *A. Rev. Ecol. Syst.*, **2**, 465–492.

[124] Harper, J.L., Clatworthy, J.N., McNaugton, I.H. and Sagar, G.R. (1961), *Evolution, Lancaster, Pa.*, **15**, 209–217.

[125] Mayer, A.M. and Poljakoff-Mayber, A. (1975), *The Germination of Seeds*, 2nd edn, Pergamon Press, Oxford.

[126] Taher, M.M. and Cooke, R.C. (1975), *New Phytol.*, **75**, 567–572.

[127] Shaw, M.W. (1974) in *The British Oak. Its History and Natural History*, (Ed.) M.G. Morris and F.H. Perring, pp. 162–181, E.W. Classey, Faringdon, England.

[128] Miles, J. (1973), *J. Ecol.*, **61**, 93–98.

[129] Miles, J. (1974), *J. Ecol.*, **62**, 675–687.

[130] Salisbury, E.J. (1942), *The Reproductive Capacity of Plants*, Bell, London.

[131] Grime, J.P. and Jeffrey, D.W. (1965), *J. Ecol.*, **53**, 621–642.

[132] Baker, H.G. (1972), *Ecology*, **53**, 997–1010.

[133] Gimingham, C.H. (1972), *Ecology of Heathlands*, Chapman and Hall, London.

[134] *Kozlowski, T.T. and Ahlgren, C.E. (Ed.) (1974), *Fire and Ecosystems*, Academic Press, New York.
      Reviews the effects of fire on vegetation in many regions, especially in North America.

[135] Daubenmire, R. (1968), *Adv. Ecol. Res.*, **5**, 209–266.

[136] Burkhardt, J.W. and Tisdale, E.W. (1976), *Ecology*, **57**, 472–484.

[137] Harper, J.L. (1961), *Symp. Soc. exp. Biol.*, **15**, 1–39.

[137a] *Harper, J.L. (1977), *Population Biology of Plants*, McGraw-Hill, London.
      An essential reference work, including sections on the seed rain, competition, grazing effects and population dynamics.

[138] Milthorpe, F.L. (1961), *Symp. Soc. exp. Biol.*, **15**, 330–335.

[139] *Rice, E.L. (1974), *Allelopathy*, Academic Press, New York.
      A useful reference work, though with some surprising omissions.

[140] Grubb, P.J., Green, H.E. and Merrifield, R.C.J. (1969), *J. Ecol.*, **57**, 175–212.

[141] Nihlgård, B. (1971), *Oikos*, **22**, 302–314.

[142] Mitchell, J. (1973), in *The Organic Resources of Scotland*, (Ed.) J. Tivy, pp. 98–108, Oliver and Boyd, Edinburgh.

[143] Byer, M.D. (1969), *Bull. Torrey Bot. Club.*, **96**, 191–201.

[144] Miles, J. (1972), *J. Ecol.*, **60**, 225–234.

[145] Ross, M.A. and Harper, J.L. (1972), *J. Ecol.*, **60**, 77–88.
[146] Black, J.N. (1958), *Aust. J. agric. Res.*, **9**, 299–318.
[147] Fricke, K. (1904), *Zentl. ges. Forstw.*, **30**, 315–325.
[148] Fabricius, L. (1927), *Forstwiss. ZentBl.*, **49**, 329–345.
[149] Fabricius, L. (1929), *Forstwiss. ZentBl.*, **51**, 477–506.
[150] Toumey, J.W. and Kienholz, R. (1931), *Bull. Sch. For. Yale Univ.* No. 30.
[151] Watt, A.S. and Fraser, G.K. (1933), *J. Ecol.*, **21**, 404–414.
[152] Korstian, C.F. and Coile, T.S. (1938), *Bull. Sch. For. Duke Univ.* No. 3.
[153] Hesselman, H. (1910), *Meddn St. SkogsförsAnst.*, **7**, 25–68.
[154] Hesselman, H. (1917), *Meddn St. SkogsförsAnst.*, **13–14**, 1221–1286.
[155] Björkman, E. and Lundeberg, G. (1971), *Stud. for. Suec.* No. 94.
[156] Muller, C.H. (1974), in *Vegetation and Environment*, (Ed.) B.R. Strain and W.D. Billings, pp. 71–85, W. Junk, The Hague.
[157] Molisch, H. (1937), *Der Einfluss einer Pflanze auf die andere—Allelopathie*, Fischer, Jena.
[158] Whittaker, R.H. (1970), in *Chemical Ecology*, (Ed.) E. Sondheimer and J.B. Simeone, pp. 43–70, Academic Press, New York.
[159] Whittaker, R.H. and Feeny, P.P. (1971), *Science*, **171**, 757–770.
[160] Nickell, L.G. (1959), *Econ. Bot.*, **13**, 281–318.
[161] Levin, D.A. (1976), *A. Rev. Ecol. Syst.* **7**, 121–159.
[162] Freeland, W.J. and Janzen, D.H. (1974), *Am. Nat.*, **108**, 269–289.
[163] Muller, C.H., Hanawalt, R.B. and McPherson, J.K. (1968), *Bull. Torrey Bot. Club*, **95**, 225–231.
[164] Handley, W.R.C. (1963), *Bull. For. Commn, Lond.*, No. 36.
[165] Robinson, R.J. (1972), *J. Ecol.*, **60**, 219–224.
[166] Webb, L.J., Tracey, J.G. and Haydock, K.P. (1967), *J. appl. Ecol.*, **4**, 13–25.
[167] Newman, E.I. and Rovira, A.D. (1975), *J. Ecol.*, **63**, 727–737.
[168] Newman, E.I. and Miller, M.H. (1977), *J. Ecol.*, **65**, 399–411.
[169] Bassalik, K. (1912), *Zeitschr. Gärungphysiol.*, **2**, 1–32.
[170] Bassalik, K. (1913), *Zeitschr. Gärungphysiol.*, **3**, 15–42.
[171] Thiel, G.A. (1927), *J. Geol.*, **35**, 647–652.
[172] Duff, R.B., Webley, D.M. and Scott, R.O. (1963), *Soil Sci.*, **95**, 105–114.
[173] Webley, D.M., Henderson, M.E. and Taylor, I.F. (1963), *J. Soil Sci.*, **14**, 102–112.
[174] Silverman, M.P. and Munoz, E.F. (1970), *Science*, **169**, 985–987.
[175] Perez-Llano, G.A. (1944), *Bot. Rev.*, **10**, 1–65.
[176] Niering, W.A. (1953), *Ecol. Monogr.*, **23**, 127–148.
[177] Winteringer, G.S. and Vestal, A.G. (1956), *Ecol. Monogr.*, **26**, 105–130.
[178] Spence, D.H.N. (1964), in *The Vegetation of Scotland*, (Ed.) J.H. Burnett, pp. 306–425, Oliver and Boyd, Edinburgh.
[179] Richardson, J.A. (1958), *J. Ecol.*, **46**, 537–546.
[180] Geiger, R. (1966), *The Climate Near the Ground*, Harvard University Press, Cambridge, Mass.
[181] Ratcliffe, D.A. (1968), *New Phytol.*, **67**, 365–439.
[182] Agnew, A.D.Q. and Haines, R.W. (1960), *Bull. Coll. Sci. Bagdad*, **5**, 41–60.
[183] Cannon, W.A. (1943), *Carnegie Inst. Wash. Publ.* No. 173.
[184] Jenny, H. (1941), *Factors of Soil Formation*, McGraw-Hill, New York.
[185] Duchaufour, P. (1968), *L'Évolution des Sols*, Masson, Paris.
[186] Woodwell, G.M. and Smith, H.H. (Ed.) (1969), *Brookhaven Symp. Biol.*, **22**.
[187] Dobben, W.H. van and Lowe-McConnell, R.H. (Ed.) (1975), *Unifying Concepts in Ecology*, W. Junk, The Hague.

[188] Goodman, D. (1975), *Quart. Rev. Biol.*, **50**, 237–266.
[189] McNaughton, S.J. (1977), *Am.Nat.*, **111**, 515–525.
[190] *Orians, G.H. (1975), in *Unifying Concepts in Ecology*, (Ed.) W.H. van Dobben and R.H. Lowe-McConnell, pp. 139–150, W. Junk, The Hague.
    An excellent analysis of the concept of stability.
[191] Holling, C.S. (1973), *A. Rev. Ecol. Syst.*, **4**, 1–23.
[192] Levin, S.A. (Ed.) (1975), *Ecosystem Analysis and Prediction*, Society for Industrial and Applied Mathematics, Philadelphia.
[193] May, R.M. (1975), *Stability and Complexity in Model Ecosystems*, 2nd edn, Princeton University Press, Princeton.
[194] *Horn, H.S. (1974), *A. Rev. Ecol. Syst.*, **5**, 25–37.
    A good discussion of certain features of secondary succession following the approach recommended by Drury and Nisbet [1].
[195] Mellinger, M.V. and McNaughton, S.J. (1975), *Ecol. Monogr.*, **45**, 161–182.
[196] Raup, H.M. (1964), *J. Ecol.*, **52** (Suppl.), 19–28.
[197] Weaver, J.E. (1954), *North American Prairie*, Johnsen Publishing Co., Lincoln, Nebraska.
[198] Medway, Lord (1972), *Biol. J. Linn. Soc.*, **4**, 117–146.
[199] Frankie, G.W., Baker, H.G. and Opler, P.A. (1974), *J. Ecol.*, **62**, 881–913.
[200] Wells, T.C.E. (1971), in *The Scientific Management of Animal and Plant Communities for Conservation*, (Ed.) E. Duffey and A.S. Watt, pp. 497–515, Blackwell Scientific Publications, Oxford.
[201] Lamb, H.H. (1965), in *The Biological Significance of Climatic Changes in Britain*, (Ed.) C.G. Johnson and L.P. Smith, pp. 3–34, Academic Press, London.
[202] Jenks, G.F. (1956), in *The Kansas Basin. Pilot Study of a Watershed*, pp. 31–41, University of Kansas, Lawrence.
[203] Korchagin, A.A. and Karpov, V.G. (1974), in *Vegetation Dynamics*, (Ed.) R. Knapp, pp. 225–231, W. Junk, The Hague.
[204] Albertson, F.W. and Tomanek, G.W. (1965), *Ecology*, **46**, 714–720.
[205] Morales, C. (1977), *Ambio*, **6**, 30–33.
[206] Gustafsson, Y. (1977), *Ambio*, **6**, 34–35.
[207] Weaver, J.E. and Albertson, F.W. (1956), *Grasslands of the Great Plains*, Johnsen Publishing Co., Lincoln, Nebraska.
[208] Thornthwaite, C.W. (1941), *Yb. U.S. Dep. Agric. 1941*, 177–187.
[209] Coupland, R.T. (1974), in *Vegetation Dynamics*, (Ed.) R. Knapp, pp. 237–241, W. Junk, The Hague.
[210] Brenchley, W.E. and Warrington, K. (1958), *The Park Grass Plots at Rothamsted 1856–1949*, Rothamsted Experimental Station, Harpenden.
[211] *Rabotnov, T.A. (1966), *Vegetatio*, **13**, 109–116.
    A fascinating example of fluctuations in temperate grassland.
[212] Miles, J. (1975), *J. Ecol.*, **63**, 891–902.
[213] Westman, W.E. (1968), *Bull. Torrey Bot. Club*, **95**, 172–186.
[214] *Watt, A.S. (1947), *J. Ecol.*, **35**, 1–22.
    The classic introduction to the ideas of cyclic change in vegetation regeneration.
[215] Warren Wilson, J. (1952), *J. Ecol.*, **40**, 249–264.
[216] Barrow, M.D., Costin, A.B. and Lake, P. (1968), *J. Ecol.*, **56**, 89–96.
[217] *Bruce, R.B. (1971), *Scott. Geogr. Mag.*, **87**, 103–115.
    Describes processes eroding vegetation on the Scottish Cairngorm Mountains.

[218] *Keatinge, T.H. (1975), *J. Ecol.*, **63**, 163–172.
      An interesting example of clonal dynamics.
[219] *Kershaw, K.A. (1962), *J. Ecol.*, **50**, 171–179.
      An intriguing example of clonal dynamics with an unknown mechanism for tiller clumping.
[220] *Watt, A.S. (1960), *J. Ecol.*, **48**, 605–629.
      A unique record of change over 21 years in a temperate grassland.
[221] Watt, A.S. (1971), in *The Scientific Management of Animal and Plant Communities for Conservation*, (Ed.) E. Duffey and A.S. Watt, pp. 137–152, Blackwell Scientific Publications, Oxford.
[222] Watt, A.S. (1975), *J. Ecol.*, **43**, 490–506.
[223] *Barclay-Estrup, P. and Gimingham, C.H. (1969), *J. Ecol.*, **57**, 737–758.
      A good account of cyclic changes in Calluna *dominated vegetation.*
[224] Tansley, A.G. (1929), *Proc. Int. Congr. Plant Sciences, 1926*, 677–686.
[225] *Haug, P.T. and Van Dyne, G.M. (1968), *Secondary succession in abandoned cultivated fields: an annotated bibliography*. Oak Ridge National Laboratory, Tennessee.
      A useful source of references.
[226] Turner, C. (1970), *Phil. Trans. R. Soc. B*, **257**, 373–440.
[227] Moe, D. (1970), *Bot. Notiser*, **123**, 61–66.
[228] Tallantire, P.A. (1972), *Norw. J. Bot.* **19**, 1–16.
[229] Andersen, S.T. (1966), *Palaeobotanist*, **15**, 117–127.
[230] Linnermark, K. (1960), *Publs. Inst. Mineral. Palaeont. Quatern. Geol.* No. 75.
[231] Miles, J. (1978), in *Institute of Terrestrial Ecology Annual Report 1977*, pp. 7–11, H.M.S.O., London.
[232] Mackereth, F.J.H. (1965), *Proc. R. Soc. B*, **161**, 295–309.
[233] Mackereth, F.J.H. (1966), *Phil. Trans. R. Soc. B*, **250**, 165–213.
[234] Ostvald, H. (1923), *Svenska Växtsoc. Sallsk. Handl.* No. 1.
[235] Godwin, H. and Conway, V.M. (1939), *J. Ecol.* **27**, 313–359.
[236] Walker, D. and Walker, P.M. (1961), *J. Ecol.* **49**, 169–185.
[237] Casparie, W.A. (1969), *Vegetatio*, **19**, 146–180.
[238] Moore, P.D. (1977), *J. Ecol.* **65**, 375–397.
[239] *Heinselman, M.L. (1970), *Ecol. Monogr.* **40**, 236–261.
      A good account of past vegetation succession and its causes on a peatland area.
[240] Ranwell, D.S. (1964), *J. Ecol.* **52**, 95–105.
[241] Lawrence, D.B. (1958), *Am. Scient.* **46**, 89–122.
[242] Sigafoos, R.S. and Hendricks, E.L. (1961), *Prof. Pap. U.S. geol. Surv.* No. 387–A.
[243] Sigafoos, R.S. and Hendricks, E.L. (1969), *Prof. Pap. U.S. geol. Surv.* No. 650–B, B89–B93.
[244] Lindroth, C.H. (1965), *Oikos*, Suppl. **6**, 1–142.
[245] Lindroth, C.H. (1970), *Endeavour*, **29**, 129–134.
[246] Crocker, R.L. and Major, J. (1955), *J. Ecol.* **43**, 427–448.
[247] Crocker, R.L. and Dickson, B.A. (1957), *J. Ecol.* **45**, 169–185.
[248] Ugolini, F.C. (1966), *Rep. Inst. Polar Stud. Ohio State Univ.* **20**, 29–72.
[249] Schoenike, R.E. (1957), *Proc. Minn. Acad. Sci.* **25**, 55–58.
[250] Cline, A.C. and Spurr, S.H. (1942) *Bull. Harv. Forest* No. 21.
[251] Spurr, S.H. (1956), *Ecol. Monogr.* **26**, 245–262.
[252] Graham, S.A. (1941), *Ecology*, **22**, 355–362.
[253] *Goodlett, J.C. (1956), *Bull. Harv. Forest* No. 25.
      Discusses the microrelief found in forests historically prone to windfall.

[254] *Jones, E.W. (1945), *New Phytol.* **44**, 130–148.
      A still useful review of regeneration in North Temperate forests.
[255] *Taylor, A.R. (1971), *J. For.* **69**, 476–480.
      Some useful statistics on the incidence of lightning fires.
[256] *Cwynar, L.C. (1977), *Can. J. Bot.* **55**, 1524–1538.
      A good case history of the incidence of lightning fires in the Algonquin Park.
[257] *Zackrisson, O. (1977), *Oikos,* **29**, 22–32.
      An account of the influence of fire in North Swedish boreal forest.
[258] Cousens, J.E. (1965), *Malay. For.* **28**, 122–128.
[259] Poore, M.E.D. (1968), *J. Ecol.* **56**, 143–196.
[260] Lloyd, M., Inger, R.F. and King, F.W. (1968), *Am. Nat.* **102**, 497–515.
[261] Harris, T.M. (1958), *J. Ecol.* **46**, 447–453.
[262] *Budowski, G. (1965), *Turrialba,* **15**, 40–42.
      Discusses four ecological strategies in tropical rain forest species.
[263] *Grime, J.P. (1977), *Am. Nat.* **111**, 1169–1194.
      Discusses ecological strategies in plants.
[264] Beckwith, S.L. (1954), *Ecol. Monogr.* **24**, 349–376.
[265] Marks, P.L. (1974), *Ecol. Monogr.* **44**, 73–88.
[266] Stephens, G.R. and Waggoner, P.E. (1970), *Bull. Conn. agric. Exp. Stn*
      No. 707.
[267] Horn, H.S. (1975), *Scient. Am.* **232**, 90–98.
[268] Shreve, F. (1942), *Bot. Rev.* **8**, 195–246.
[269] Muller, C.H. (1952), *Bull. Torrey Bot. Club,* **79**, 296–309.
[270] Babb, T.A. and Bliss, L.C. (1974), *J. appl. Ecol.* **11**, 549–562.
[271] Levin, M.H. (1966), *Am. Midl. Nat.* **75**, 101–131.
[272] *Cremer, K.W. and Mount, A.B. (1965), *Aust. J. Bot.* **13**, 303–322.
      A study of succession in felled and burnt Eucalyptus forest.
[273] *Kellman, M.C. (1970), *Secondary Plant Succession in Tropical Montane
      Mindinao.* Dept. of Biogeography and Geomorphology Publ. BG/2,
      Research School of Pacific Studies, Australian National University.
      *A valuable study of factors influencing secondary succession in tropical
      forest.*
[274] *Webb, L.J., Tracey, J.G. and Williams, W.T. (1972), *J. Ecol.* **60**, 675–695.
      *An experimental study of secondary succession in sub-tropical rain forest,
      with a highly recommended discussion.*
[275] *Farnworth, E.G. and Golley, F.B. (1974), *Fragile Ecosystems. Evaluation
      of Research and Applications in the Neotropics.* Springer, Berlin and New
      York.
      Contains a summary of secondary succession in neotropical vegetation.
[276] Muller, C.H. (1940), *Ecology,* **21**, 206–212.
[277] Judd, B.I. (1940), *J. Am. Soc. Agron.* **32**, 330–336.
[278] Davis, R.M. and Cantlon, J.E. (1969), *Bull. Torrey Bot. Club,* **96**, 660–673.
[279] Kramer, F. (1933), *Tectona,* **26**, 156–185.
[280] Kellman, M.C. (1969), *Syesis,* **2**, 201–212.
[281] Knight, D.H. (1975), *Ecol. Monogr.* **45**, 259–284.
[282] Williamson, G.B. (1975), *Ecology,* **56**, 727–731.
[283] Coile, T.S. (1940), *Bull. Sch. For. Duke Univ.* No. 5.
[284] McQuilkin, W.E. (1940), *Ecology,* **21**, 135–147.
[285] Miles, J. (1973), *J. Ecol.* **61**, 399–412.
[286] *Williams, W.T., Lance, G.N., Webb, L.J., Tracey, J.G. and Dale, M.B.
      (1969), *J. Ecol.* **57**, 515–535.
      *An antecedent study to [274], with which it should be read. Shows the
      danger of drawing conclusions about succession based on short periods of
      study only.*

[287] *Niering, W.A. and Goodwin, R.H. (1974), *Ecology*, **55**, 784–795.
  Includes excellent brief reviews of the dynamics of old-field succession and of the occurrence of stable vegetation that resists tree invasion.
[288] Wells, B.W. (1937), *J. Elisha Mitchell scient. Soc.*, **53**, 1–26.
[289] Rice, E.L. (1972), *Am. J. Bot.* **59**, 752–755.
[290] *Franklin, J.F. and Dyrness, C.T. (1969), *Res. Pap. Pac. N.W. For. & Range Exp. Stn*, No. PNW-80.
  Includes brief accounts of secondary succession, especially of forest, in Oregon and Washinton, including the formation of stable, shrub-dominated vegetation.
[291] Symington, C.F. (1933), *Malay. For.* **2**, 107–117.
[292] Gliessman, S.R. and Muller, C.H. (1971), *Madroño*, **21**, 299–304.
[293] Glass, A.D.M. (1976), *Can. J. Bot.* **54**, 2440–2444.
[294] Cottam, G. and Wilson, H.G. (1966), *Ecology*, **47**, 88–96.
[295] Horn, H.S. (1975), in *Ecology and Evolution of Communities*, (Ed.) M.L. Cody and J.M. Diamond, pp. 196–211, Harvard University Press, Cambridge.
[296] Horn, H.S. (1976), in *Theoretical Ecology. Principles and Applications*, (Ed.) R.M. May, pp. 187–204, Saunders, Philadelphia.
[297] Huffaker, C.B. and Kennett, C.E. (1959), *J. Range Mgmt*, **12**, 69–82.
[298] Cantlon, J.E. (1969), *Brookhaven Symp. Biol.* **22**, 197–205.
[299] Piemeisel, R.L. (1945), *Circ. U.S. Dep. Agric.* No. 739.
[300] Jones, L.I. (1967), *Tech. Bull. Welsh Pl. Breed. Stn.* No. 2.
[301] Miles, J. (1979), in *Upland Land Use in England and Wales*, (Ed.) O.W. Heal, Countryside Commission, Publication CCP111. (In press)
[302] Yapp, W.B. (1953), *NWest. Nat.* **24**, 190–207, 370–383.
[303] *Connell, J.H. and Slatyer, R.O. (1977), *Am. Nat.* **111**, 1119–1144.
  An up-to-date review and discussion of models of succession in vegetation.

Index

78

inhibition model, 16
initiation, 21–22
in open water, 34
models, 13–16, 60, 65
on abandoned fields, 14, 47, 57, 59
on glacial debris, 20, 51–53
on sand dunes, 20
primary, 14, 20, 21, 22, 36, 47–53, 66
probabilistic, 14, 60
properties, 16–19
secondary, 14, 22, 36, 37, 54–60, 66
Succiza pratensis, 37
Sugar maple, see *Acer saccharum*
Surtsey, 26
Swamping, 50

Taiga, 38
Teucrium scorodonia, 15
Texas, 23
Trifolium hybridum, 39
Trifolium pratense, 39
Trifolium subterraneum, 32
Trichophorum cespitosum, 15
Tsuga spp., 55
Tsuga heterophylla, 51
Tsuga mertensiana, 51
Tundra, 54, 55, 58

Typha latifolia, 51

Ulex europaeus, 29
United States, northeastern, 45
southwestern, 19
western, 31

Vaccinium myrtillus, 15, 63
Vegetation, classification, 9, 11, 65
flux, 11, 14, 21, 26, 36, 54, 55
nature of, 7–13, 65
properties, 9–11
variation in space, 7–13, 65
Vegetation change, evidence for, 19, 20
Vegetative spread, 7, 25, 41
Veronica officinalis, 15
Veronica serpyllifolia, 15
Vicia cracca, 39
Volcanic eruptions, 21, 25, 26

Wales, North, 24
Wave-regeneration, 45, 46
Wind-fall (wind-throw), 10, 22, 36, 54, 58

Zerna erecta, 37
Zerna inermis, 39

BELMONT COLLEGE LIBRARY